Emerging Globalities and Civilizational Perspectives

This series documents the range of emerging globalities in the 21st century at the national, transnational and trans-civilizational levels of analysis. "Globality" refers to a global condition where people located at any point on Earth are aware of being part of the world as a whole---the world as a single interacting entity. Social interactions occur among actors belonging to different societies, different social strata and different cultural traditions so that the condition of "globality" is experienced in many different ways.

Examples of emerging globalities are social movements generated from the unfulfilled promises of neoliberalism and feelings of discrimination and marginalization of lower social strata; cultural otherization or the blaming of economic problems of certain geographical areas on a low level of cultural development; insecurities generated by technological risks, epidemics, and global terrorism; uncertainties generated by processes of transnational governance, outsourcing, unbalanced trade and massive migrations; biology-machine interfaces and impacts of non-human organisms and technologies on human consciousness and action; long-term threats of global warming, climate change and depletion of bio-diversity; increasing exploitation and marginalization of less industrialized regions.

We state that globalization entails encounters and often clashes among people and nations of different civilizational traditions. Hence, one of the exploratory questions of these volumes will be the extent to which negative or problematic globalities are reactions to failed promises and unrealized ideals of civilizational and national traditions and/or perhaps attempts to revive those traditions. Our notion of civilizational tradition takes inspiration from the classical works of Spengler and Toynbee, Benjamin Nelson, Vytautas Kavolis, Roland Robertson, Johann P. Arnason, Jeremy Smith, and others; a tradition which is in sharp contrast with the civilizationism recently promoted by authoritarian leaders with hegemonic ambitions. The volumes in this series aim to extend the inter-civilizational focus of classical civilizational thinkers from the analysis of the origins and development of civilizations to the fostering of contemporary inter-civilizational dialogues; the intent is to facilitate an international rapprochement in the contemporary atmosphere of global conflicts.

The volumes will reflect the diversity of theoretical perspectives and captures some of the novel thinking in social sciences, economics and humanities on intra- and inter-societal processes; the attention to novel thinking will extend to emerging policy formulations in dealing with threats, risks, insecurities and inequities and to strategic thinking for a sustainable global future. The historical perspective will also be an important component of analysis together with the avoidance of West-centric perspectives. The intended readership of this series is not just an academic audience but also policy decision-makers and the public at large; accessibility of language and clarity of discourse will be a key concern in the preparation of these volumes.

Joy Y. Zhang • Kathleen M. Vogel
Sonia Ben Ouagrham-Gormley • Ruipeng Lei
Editors

BioGovernance Commons and China's Role in Global Risk Governance

Springer

Editors
Joy Y. Zhang
Centre for Global Science and
Epistemic Justice
University of Kent
Canterbury, UK

Sonia Ben Ouagrham-Gormley
Schar School of Government & Policy
George Mason University
Arlington, VA, USA

Kathleen M. Vogel
Arizona State University
Tempe, AZ, USA

Ruipeng Lei
Advanced Inst. of Hum. & Soc. Sc.
University of Electronic Sc & Tech
Chengdu, Sichuan, China

ISSN 2731-0620 ISSN 2731-0639 (electronic)
Emerging Globalities and Civilizational Perspectives
ISBN 978-3-032-32197-8 ISBN 978-3-032-32198-5 (eBook)
https://doi.org/10.1007/978-3-032-32198-5

This Springer imprint is published by the registered company Springer Nature Switzerland AG
The registered company address is: Gewerbestrasse 11, 6330 Cham, Switzerland

Acknowledgments Vogel, Ben Ouagrham-Gormley, and Zhang acknowledge support from the U.S. National Science Foundation (grant no. 2112906). Ruipeng Lei acknowledges support from the National Key R&D Program of China (grant no. 2024YFA-9th-17200). Zhang also acknowledges financial and administrative support from the Centre for Global Science and Epistemic Justice at the University of Kent.

We thank Professor Ino Rossi, Shinjini Chatterjee and the editorial team at Springer for their support and commitment in bringing this volume to fruition.

We extend our deepest gratitude to the hundreds of participants who have contributed to BioGovernance Commons events around the world through their insights, critiques, and support.

Contents

Chapter 1
The Creation of BioGovernance Commons

Joy Y. Zhang, Kathleen M. Vogel, Sonia Ben Ouagrham-Gormley, and Ruipeng Lei

Abstract This chapter introduces the concept, practice, and evolving community of the BioGovernance Commons, a transnational experiment in sustaining dialogue on life science governance amid intensifying geopolitical tension. Originating in 2021 as a 1-year initiative connecting scholars and practitioners across China, the United Kingdom, and the United States, the BioGovernance Commons was designed to promote cross-border dialogue at a time when scientific exchange was increasingly strained by mistrust, censorship, ideological polarisation, and institutional closure. Drawing on the idea of the 'commons' not as a static shared resource but as an ongoing practice of 'commoning', the chapter argues that trustworthy transnational coordination in science cannot rest on assumptions of universalism, consensus, or uniformity. Instead, it requires sustained relational work through which differences are rendered intelligible, disagreements are negotiated, and mutual understanding is cultivated without erasing asymmetry or conflict. The chapter also reflects on the making of the volume itself, showing how challenges of collaboration, authorship, academic form, and self-censorship expose both the fragility and the necessity of commoning. Overall, the BioGovernance Commons is positioned as both an empirical case and a normative proposition for reimagining global biogovernance in fractured worlds.

Keywords Global governance · Commoning · Science diplomacy · Geopolitics of science · Transnational collaboration · Epistemic pluralism

---

J. Y. Zhang (✉)
Centre for Global Science and Epistemic Justice, University of Kent, Canterbury, UK
e-mail: yz203@kent.ac.uk

K. M. Vogel
Arizona State University, Tempe, AZ, USA

S. Ben Ouagrham-Gormley
George Mason University, Fairfax, VA, USA

R. Lei
University of Electronic Science and Technology of China, Chengdu, China

J. Y. Zhang et al. (eds.), *BioGovernance Commons and China's Role in Global Risk Governance*, Emerging Globalities and Civilizational Perspectives,
https://doi.org/10.1007/978-3-032-32198-5_1

The Origin of an Experiment

This book grew out of our experimental effort to create a BioGovernance Commons, a transnational experiment in sustaining dialogue on life science governance amid intensifying geopolitical tension. What began as a 1-year initiative aimed at fostering productive and sustainable dialogue among life science communities in the United States, the United Kingdom, and China has since evolved into a connective thread linking an expanding, interdisciplinary group into a transnational community. Together, this community seeks to (re)build confidence in effective transnational scientific and regulatory collaboration amid geopolitical tensions—without enforcing uniformity, and in ways that remain attentive to difference, diversity, and our shared planetary future.

What prompted us to initiate the BioGovernance Commons experiment was a convergence of geopolitical tensions and institutional developments in 2021. Over the past two decades, sustained government investment in science has propelled China from the world's ninth-largest biomedical sector in 2006 to fourth place by 2015, and by 2021 China had reached second place globally—after the United States—in biological sciences, measured primarily by publication output in major bibliometric databases, with supporting evidence from citation-based metrics (see Ben Ouagrham-Gormley and Vogel 2020a, 2020b; Vogel and Ben Ouagrham-Gormley 2022; Zhang 2021b; Trimble 2017). China's position in global science alone would warrant serious and sustained engagement with Chinese scientific communities. Yet, a series of compounding circumstances has made such engagement not only desirable but increasingly urgent.

Alongside the rapid expansion of scientific capacity and output, China has also faced a succession of high-profile research mishaps and cases of scientific misconduct, as well as longstanding challenges related to transparency and regulatory enforcement. These issues have cast a persistent shadow over the credibility of parts of China's scientific ecosystem. At the same time, negotiations to update the EU–China research collaboration roadmap stalled (Kelly 2021); academic censorship in China tightened further in the wake of the COVID-19 pandemic (Zhang and Barr 2021); and the United States' China Initiative, a programme aimed at countering economic espionage that, in practice, intensified scrutiny of Chinese and Chinese-origin researchers and raised concerns about racial profiling and constraints on international collaboration, has come to symbolise an intensifying global ideological divide (Lewis 2020). For many Chinese academics, the combined effects of heightened domestic censorship and growing suspicion from foreign media and regulatory authorities have increasingly pushed the prospect of meaningful global engagement into a state of impasse.

It was against this backdrop of scientific prominence, political friction, and institutional closure that the BioGovernance Commons was conceived—as an attempt to reopen channels of dialogue at a moment when they were rapidly narrowing. More specifically, we wanted to break free of harmful stereotypes and overgeneralizations, but also to reveal some of the hidden prejudices on both sides that can prevent fruitful collaborations. We decided to host a trusted forum where researchers inside and outside of China could meet regularly, and as our motto stated, to 'share perspectives on shared challenges.' We named our initiative 'BioGovernance Commons.' We follow Elinor Ostrom's (1990) reasoning and consider the word 'commons' as a verb. That is, we take the view that any form of community-building and dialogue-sharing is achieved through continuous practices of *commoning*: acts of support, conflict, negotiation, and experimentation that are needed to shape a transnational knowledge system with sustainable relations among individuals.

We began with monthly online meetings among a small group of trusted colleagues working on life-science governance across the UK, the USA, and China, who had collaborated before or were familiar with one another's work. Each meeting consists of 15–20 participants from Europe, Asia, and North America with related social science or natural science backgrounds. Our discussions were organised around specific readings and presentations by experts from Asia, Europe, or North America. We also kept in touch with many of the participants on science governance issues outside of these meetings through social media (WeChat) groups and emails. As the conversations deepened, the circle of participants gradually expanded. While most of our meetings involve a couple of dozen participants to allow for in-depth conversation, some of our events—such as the discussion on Chinese entrepreneur Jiankui He's planned DMD clinical experiment—had more than 100 participants (see Zhang et al. 2023a, 2023b). Not all participants attended every meeting, although we did have a few regulars, including the editors of the volume and several contributors, such as Lu Gao, Haidan Chen, and Daniela Caterina. Regardless of the meeting size and issue under discussion, these exchanges consistently centred on a shared set of questions: how should we understand the logics of China's research culture? Which 'differences' are negotiable, and which demand respect? Where do blind spots persist—both within China and beyond? What are the sources of frustration, internally and in external perceptions of Chinese science? And, critically, how might these frictions be mitigated? Over time, this sustained engagement has cultivated a small but growing community grounded in mutual understanding and trust, demonstrating the possibility of meaningful dialogue on life-science governance, including biosecurity, genome editing, and cross-border data sharing, even amid widening geopolitical constraints—such as tightening research security policies, data-sharing restrictions, visa and mobility barriers, and increased scrutiny from funders, regulators, and media, particularly since the late 2010s. Although the initiative was originally envisaged as a 1-year experiment, the easing of COVID-19 restrictions enabled us to extend these exchanges into in-person meetings, held in London (2022), Chengdu (2023), and Dalian (2025).

While unplanned, the BioGovernance Commons initiative corresponds to the ethos of the Emerging Globalities and Civilizational Perspectives series. BioGovernance Commons takes emerging globality seriously. Our initiative echoes the series' critique of 'the cultural otherization or the blaming of economic problems of certain geographical areas on a low level of cultural development;

insecurities generated by technological risks; and uncertainties generated by processes of transnational governance'.

The chapters included here are authored by individuals who have contributed to the BioGovernance Commons initiative at different stages. This book is not a straightforward documentary of the discussions that took place. Rather, it demonstrates how the idea and practice of the Biogovernance Commons have been, and could be, mobilised internationally, while shedding new light on the socio-political organisation and dynamics of the Chinese life sciences in particular, and on the forging of new dialogues and collaborations more broadly. Before explaining the making and structure of this book, it is necessary to clarify what the 'BioGovernance Commons' is, and how it operates simultaneously as a concept (its core values), a practice (its modes of organisation), and a community (its evolving impacts).

BioGovernance Commons as a Concept

The dialogical impasse over trustworthy transnational science and regulatory coordination on emerging biotechnologies, in our view, fundamentally requires a 'relational thinking—that of the relationship between place, knowledge and power' (Roy 2016, 207). On the one hand, we need to qualify the old saying 'science is universal'. Is science 'universal'? Conventionally, from the perspective of the Global South, such a claim itself is an assertion of Eurocentric epistemologies. For science has never been an exercise of disinterested fact gathering, but has always been 'an uneven distribution of epistemic authority', which imposes order over knowledge and those who know (Hamilton et al. 2017: 615). But on the other hand, there is a heightened necessity to make broader and more sustainable coordination in the governance of science, because in terms of its *impact*, we argue that the global outreach of science cannot be emphasised enough. As political scientist Brian Salter (2008: 156) rightly argued, in contemporary global innovation, 'no state can afford to be a political island'. Science governance would be much easier if only the social-ethical consequences of controversial practices or disputable risk-benefit judgements could be confined to particular geographical boundaries, or to a neat West-East divide (Bharadwaj 2012). But as we have seen in the life sciences, from experimental trials of regenerative medicine to reproductive human genome editing (CRISPR Baby scandal, see Liu et al. 2024, Ouagrham-Gormley and Vogel forthcoming), the exploitation of scientific viability in one corner of the world often profoundly reorients the perception of technological imperatives and associated risk calculations around the world. Thus, to maximise the benefits and to minimise the risk of emerging science in aiding our collective pursuit of a good life, we need to acknowledge and effectively respond to the universal potential of science's impact. For both contemporary science and its governance, we need what Sandra Mitchell (2009: 16) called 'integrative pluralism', in which a plurality of perspectives can be integrated to reflect the dynamic and evolving character of knowledge.

Put differently, science is neither universal nor does it automatically constitute a 'commons'. Rather, it is a collective socio-political practice: an evolving set of social relations that is always a work in progress and that requires deliberate, attentive acts of 'commoning'. It was for this reason that we turned to Elinor Ostrom's foundational work on the commons. Ostrom conceptualised commons not simply as shared resources, but as configurations linking resources, those entitled to use them, and the governance arrangements designed to manage them. Her work demonstrated how well-informed individuals can undertake collective action to protect shared resources without causing irreparable degradation, and without necessarily relying on state intervention or private property regimes (Ostrom 1990). Subsequent scholarship on the commons has further emphasised that institutions governing shared resources are not merely sets of formal rules or laws. They are embedded in social and ecological relationships, shaped by power asymmetries and inequities that influence whether, how, and to what extent individuals choose to participate (Nightingale 2011; Singleton 2017).

The transformative power of commoning in relation to contemporary China was first observed through empirical studies of food movements (Zhang and Barr 2019). Commons are not merely resources that are owned or accessed collectively; they also constitute a nexus of place, intricate social relations, and collectively defined norms and interests. Most crucially for this book, commoning is not primarily about achieving consensus. Rather, it provides the ground upon which disagreement can be articulated, negotiated, and sustained in meaningful ways.

Put differently, the BioGovernance Commons treats the creation of a 'commons' as essential to our shared future, while retaining the humility to recognise that such a commons is neither given nor something that can be designed or imposed from a single perspective. As a concept, it reconfigures both the means and ends of transnational encounters. Rather than seeking agreement or other forms of unified commitment, we understand the breaking of impasse to lie in enabling different perspectives to make better sense of their differences. Instead of treating convergence of ideas as the primary objective, we view it as a possible but secondary by-product. The BioGovernance Commons also invites us to reinterpret forces conventionally seen as divisive (such as difference and conflict) as potential sources of complementary strength and as openings for coordination. What we foreground, instead, is the capacity to render differences intelligible to one another, an orientation we regard as more empowering for navigating complex and contested scientific and political terrains.

BioGovernance Commons as a Practice

Shortly after the World Health Organization visited the Wuhan Institute of Virology on 3 February, a reporter from an international newspaper asked me whether China's censorship would present an 'insurmountable' barrier to reaching an impartial conclusion on COVID-19's origins. At the same time, my phone displayed headlines from a nationalist

> Chinese newspaper proclaiming that the mainstream media's 'smearing' of the investigation was rooted in denial of the 'truth'. They came from a portal managed by China's scientific societies.
>
> This distrust is part of an intensifying pattern. Without constructive engagement between scientists, it will get worse…Those I spoke to early in 2020 in Beijing, Xi'an, and Wuhan had anticipated censorship and frowned upon it.
>
> Several months later, as COVID science became increasingly politicised, several researchers described the 'necessity' of government controls over information. 'Western media is always xiashuo!' one geneticist told me—a sentiment I've heard often. Xiashuo, literally 'talking blindly', means speaking nonsense or spreading rumours. I reminded her that over-generalizing about 'Western media' was just as biased as Western over-generalization about Chinese science. I described how, during the pandemic, journals such as The Lancet, Science, and Nature have consistently advocated a fair appreciation of Chinese science. She rebuked me with quotes from mainstream media describing China's research and development as a nefarious 'rising red moon' that would 'always be bad at bioethics' (Zhang 2021a).

The passage above was written by one of us (Zhang) for *Nature*, calling for global academic solidarity to defend knowledge exchange and sustain sensible, constructive dialogue, just a few months before we convened the first BioGovernance Commons meeting. At the time, amid deepening mutual scepticism between China and the West, we were acutely aware that engagement with China could, to paraphrase Dickens, usher in either 'the spring of hope' or 'the winter of despair'. It was with this ambivalence firmly in mind that the four of us, editors of this volume, began by reaching out to a small group of colleagues whom we believed possessed not only relevant expertise, but also the intellectual openness, reflexivity, and relational sensitivity required for such a conversation.

This prudence proved necessary. When our participants (bioethicists, life scientists, historians of science, and science and technology scholars) met online for the first couple of BioGovernance Commons sessions, even within this small circle, we became aware that words once considered mundane had been politicised through ongoing media and policy debates. Such terms were open to divergent interpretations and, at times, prompted scepticism about speakers' intentions. As a result, our discussions frequently took pauses to invite participants to clarify what they meant, and equally importantly, what they did not mean. For example, was referring to China as part of the 'Global South' intended as a derogatory classification, or as an empathetic or strategic positioning? These moments of pausing and unpacking different interpretations became an essential part of building mutual understanding, rather than an obstacle to it. It also underpins the point that simple acts of 'respect' and 'equality' may not bring cohesion either, for it may legitimise distance or worse, indifference. Thus, we draw attention to the inherently subjective process of 'making common' and urge participants to recognise and actively address asymmetries in knowledge and perspective.

In fact, after the initial phase, we had frank and critical discussions on thorny issues, such as genetic data-sharing, biosecurity, enforceability of institutional ethical review, and commercialisation of CRISPR technologies. We've detailed some of the key lessons learned in a perspective piece in the journal *Issues in Science and Technology* (see Zhang et al. 2022).

Over the initial 12 months, the initiative received support from scientific and social science professionals in Brazil, China, Italy, Greece, the UK, and the USA. Between August and September 2022, we organised what were originally planned as two concluding forums: one a US–China dialogue on 'How to Build Better Global Governance of Science', and the other an EU–China dialogue on 'How to Make "the China Story" More Informative' (see the programmes on the GSEJ blogs, links included at the end of the references list). These online forums brought together around 40–60 participants per session, including scientists, social scientists, policy practitioners, and science communicators from both sides. Discussions focused on improving mutual intelligibility across scientific and policy cultures, addressing asymmetries in knowledge production and representation, and exploring practical pathways for more reflexive and inclusive forms of global science governance and communication. In line with our broader discussions, we understand making the China story of science and innovation more informative as a way of promoting global understanding of China. We emphasise that 'understanding' does not necessarily lead to agreement; rather, it is a precondition for constructive and accountable collaboration.

BioGovernance Commons as a Community

Perhaps the most telling indicator of the success of our original 'experiment' in creating a biogovernance commons is that the network we brought together did not disintegrate. Instead, BioGovernance Commons has continued to function as a trusted convening space and a shared point of reference, even amid periods of heightened transnational scepticism and political tension.

The most consequential example of this impact was the role BioGovernance Commons played in supporting Chinese scientists' efforts to halt further activities by biophysicist Jiankui He. Jiankui He is the infamous scientist responsible for the world's first instance of heritable human genome editing, which resulted in the birth of three girls in 2018 (Cyranoski 2018). These actions violated both Chinese regulations and internationally recognised ethical red lines, and He was sentenced to 3 years in prison. However, upon his release, He, with no clinical training, publicly announced plans to embark on experimental gene therapy for Duchenne muscular dystrophy (DMD). This proposed venture was marked by limited scientific credibility, an absence of clear regulatory oversight, and the worrying fact that patient families had already been mobilised. Compounding these risks, Chinese censorship practices—particularly the removal of contentious or controversial discussions from public view—meant that there was no sustained public record or scrutiny of his new activities.

While many Chinese academics were deeply concerned about the potential harm to children with Duchenne muscular dystrophy and the prospect of another medical scandal, there were few viable avenues through which they could publicly voice their concerns. At the same time, Jiankui He, through his egregious malpractice, continued to generate sensational headlines and, in the process, became the face of a reductive and distorted image of 'Chinese science'. As domestic media hesitated to report a scandal in the making, foreign media coverage often appeared unsympathetic to the broader Chinese scientific community, further entrenching polarised narratives. It was precisely under these conditions that transnational solidarity within the global scientific community could, and did, make a meaningful difference.

On 11 February 2023, we used the BioGovernance Commons platform to convene an open discussion on CRISPR technologies with and for Chinese academics. The event, titled *Looking Back into the Future* (see Zhang et al. 2023a, 2023b and Appendix 2), brought together approximately 110 participants from 13 countries, including at least 70 Chinese academics representing 12 provincial jurisdictions. Notably, this was the first time that Dr. He agreed to engage publicly with Chinese bioethicists and fellow CRISPR scientists. The discussion revealed that there was little substantive basis behind the ambitious DMD gene therapy claims that He had been promoting to prospective patients worldwide. Participants expressed serious concern that, if left unchecked, his activities could once again place a highly vulnerable population at risk.

The event had immediate effects. Chinese media outlets publicly questioned He's DMD research for the first time, and within 36 h of the discussion, He discontinued his online marketing to patients. Following the release of our conference report, coverage by Chinese and international media ensured that a fuller account of the event entered the public domain. This, in turn, enabled Chinese bioethicists, legal scholars, and scientists to issue a consensus statement in the early hours of 5 March 2023 (Frontiers of Bioethics Law 2023). The statement condemned He's actions, urged Chinese authorities to be more accountable in their administrative responsibilities, and explicitly protested against the censorship and secrecy surrounding the government's response. For many scientists and regulators worldwide, this intervention, released on the eve of the Third International Summit on Human Genome Editing, was both timely and significant, breaking a prolonged silence from China since He's sentencing in 2019. The timing of the statement was particularly striking. It was released on the opening day of China's National Congress, a period typically associated with heightened censorship and intolerance of dissent. In this context, the publication of the consensus statement was arguably more remarkable than it might otherwise appear.

BioGovernance Commons helped prevent another spurious clinical trial in gene therapy, while also providing a trusted space for international solidarity and open, public discussion of deeply controversial issues in biomedical research—even when political, institutional, and epistemic conditions make such openness difficult to sustain.

The Making of this Book

In editing this volume, we pursued two sets of ambitions that may at first appear paradoxical. First, we aim for the book to be both informative and provocative. On the one hand, it offers readers empirically grounded insights into how the organisation and governance of science in China have evolved with the global socio-political climate. On the other hand, it intentionally creates space for contributors' individual interpretations, reflections, and aspirations—perspectives that are not always easily accommodated within conventional peer-reviewed journal formats. These personal, informed, yet deliberately less 'filtered' reflections from scholars working at the frontlines of global governance research are intended not only to inform, but also to provoke further collective reflection and differentiated forms of action in the development of future international ecosystems for responsible science and governance.

The second, related ambition is for the book to both capture a particular historical moment and speak beyond it. Specifically, we seek to document China's evolving position in the world, along with the questions, uncertainties, and perplexities this position raises—both globally and within contemporary Chinese academic communities themselves. At the same time, we hope the volume will endure beyond its immediate geopolitical context. By foregrounding the differentiated rationales through which contributors engage with the idea of a BioGovernance Commons—its possibilities as well as its limitations—the book aims to offer more durable insights into how difference can be negotiated productively, and how collaboration can be sustained across epistemic, cultural, and political divides that transcend any single socio-political moment.

We are pleased to bring together this manuscript, which includes not only the small circle of academics who made the original 1-year BioGovernance Commons project possible, but also a broader group of contributors who have engaged with the initiative at different stages, in varying capacities, and from diverse academic and socio-political perspectives. The contributors span a wide range of disciplines—including the history of science, anthropology, sociology, political science, law, and security studies—which, while distinct, can be broadly situated within the interdisciplinary terrain of science and technology studies. They are also at different career stages, yet all are actively involved in the transnational governance of emerging life sciences and bring first-hand experience of the geopolitical and transcultural challenges that shape this field.

The making of this book has not been an easy journey. The process itself bears witness both to the vulnerability of *commoning* in a world deeply imbricated with the logics of narrowly defined productivity and competition, and to its essential role in enabling us to explore alternatives for a more constructive, coordinated, and shared future.

Our first set of challenges was practical and structural. Although the idea of collectively producing this volume emerged in June 2022, the writing process was repeatedly delayed, for reasons that were largely symptomatic of broader transformations in academia since the COVID-19 pandemic. Across China and elsewhere, academic workloads have increased dramatically—often doubling or even tripling—while funding cuts, institutional restructuring, and intensified performance metrics have become commonplace. In China, the sudden reversal of stringent COVID-19 testing and lockdown measures in December 2022, implemented without adequate public health preparation, led to widespread infection across cities and universities. Arguably, it was as challenging a time for China-based researchers as the beginning of COVID. At the same time, higher education sectors in many Western countries have been contracting, with growing precarity and threats to job security. Against this backdrop, a few contributors, both in China and in the West, withdrew from the book project. While committed to the ethos of the book, they noted that contributing to an edited volume carries limited institutional recognition, particularly in promotion and evaluation processes. Some candidly told us they would have participated had the project taken the form of a special journal issue. Yet journal formats, with their established conventions and constraints, leave little room for the personalised, reflexive, and sometimes unfiltered reflections that this book seeks to encourage and preserve. It was precisely for this reason that we chose the freedom of the book form. Even so, the pace of writing proved slower than anticipated, and it was not until 2023 that the first full drafts began to come together.

A second challenge then emerged—one that may appear counterintuitive, yet proved deeply revealing: how to incorporate personal reflection within academic writing. This difficulty was particularly acute for some contributors based in China. Despite being generous and insightful in sharing personal perspectives during BioGovernance Commons discussions (which were intentionally not recorded), several later confided, most directly to one of us (Joy), that they struggled to write such reflections on paper. 'Because it would no longer be "academic", wouldn't it?' one contributor asked.

We consider this hesitation significant rather than incidental, as it exemplifies precisely the academic–political entanglements that the BioGovernance Commons initiative seeks to surface and, where possible, to reconfigure. The question of what counts as 'being academic' is crucial here. On the one hand, years of government censorship in China, combined with online harassment and trolling, a phenomena not unique to China, have fostered a pervasive culture of self-censorship, or more accurately, strategies of self-preservation (Zhang 2015; Zhang and Barr 2021). Many academics refrain from writing commentaries for foreign outlets. It is widely known that Chinese academics often seek institutional approval before accepting media interviews, whether domestic or international. Such practices function not only as formal requirements (especially in relation to non-Chinese media), but also as mechanisms of institutional protection should public controversy arise.

In fact, just as the book was about to enter the production stage, one of the original contributors requested to withdraw her chapter. She is an expert in global health security governance, with experience working in both China and the United States, and a member of the National Expert Pool for Global Health Governance. Her chapter examined how international law and global governance systems attempt, often imperfectly, to manage global health security in the aftermath of COVID-19.

She explained that she was concerned her ideas were 'not sufficiently mature' (*bugou chengshu*) and might be of 'potential (political) inappropriateness' (*keneng buheshi*). More importantly, she felt that, for her own protection, publication would require full institutional approval and a formal institutional 'declaration' (*shengming*) that would shield her from future political accountability.

This is a highly unusual request for several reasons. To begin with, she is not only a leading expert on the topic in China, but her chapter has also been developed and revised over a period of 6 months. Academically, it is far from 'immature.' On the contrary, it offers a nuanced account of how China's national approach emphasises immediate threat mitigation, rapid emergency response, and collective mobilisation, in contrast to Western frameworks that prioritise prevention, preparedness, and institutional resilience. By situating these differences within broader cultural and historical contexts, the chapter moves beyond simplistic geopolitical explanations and instead highlights deeper 'transcultural' variations in how security itself is understood.

Her concern about political appropriateness may seem equally perplexing. The chapter presents an evidence-based critique of global governance, arguing that global health governance cannot be understood solely as a state-to-state issue but rather as a complex web of interactions both between and within state and non-state actors. It identifies two major governance dilemmas: first, the challenge of balancing prevention across human, animal, and environmental domains under conditions of limited resources and constrained sovereignty, and second, the question of whether countries will share pathogens when the benefits (such as vaccines) are distributed unequally. The chapter concludes by proposing practical steps that could be taken even before new treaties are fully implemented, including strengthening global financing, improving national capacity assessments, and expanding support for developing countries. The manuscript was also reviewed by Chinese academics at both the draft stage and as part of the full volume, and none of the reviewers—well attuned to censorship sensitivities—flagged any concerns. However, as the author later explained, her primary worry was not the *present* but the *future*: that shifting political conditions might lead to her work being read in an unfavourable light or her intentions being misinterpreted as politically inappropriate. For this reason, she felt unable to proceed without institutional approval, or, more precisely, institutional guarantees of protection against future repercussions. She also candidly noted that she has published very little in recent years for the same reason. When Joy suggested that she might publish the piece independently, without institutional affiliation, she declined, responding: 'Then it's of no use to me. Why would I publish at all?'

At one level, this kind of 'pre-compliance' with anticipated censorship might be criticised as a failure to speak truth to power or to defend academic freedom. However, it is important not to reduce the situation to a question of individual courage. Rather, it reveals the depth of the moral dilemmas that Chinese academics must navigate, as well as the profound anxieties that shape their decisions. Notably, this professor was not someone who simply 'plays safe': she was keen to contribute to the BioGovernance Commons project, submitted a full draft despite initial time constraints, and carefully revised it in response to reviewers' comments.

While the 'possibility' of political inappropriateness may appear abstract or even exaggerated, for individual academics, the risks associated with shifting political climates, censorship, and institutional accountability are deeply tangible. Weighed against the potential consequences for her long-term employment and her roles on national and regional expert committees, her decision becomes, if not unproblematic, at least understandable. In this sense, the dynamics of censorship create a paradoxical environment in which those with the most relevant expertise and experience may be the least able, or willing, to share their carefully developed views, for fear of political misattribution and threats to their professional security.

This is a political reality that the BioGovernance Commons initiative has had to confront, and it only underlines the urgency and value of keeping arenas open in which relatively candid discussion remains possible. While the author remains in contact with many colleagues within BioGovernance Commons and is still committed to international dialogue, withdrawals such as this hinder future collaboration in general, as they prevent new collaborators—especially the younger generation of researchers—from understanding China and deny them the opportunity to build trust with Chinese experts such as herself.

However, at least in theory, academic channels are still seen as distinct from the civic-political sphere and still offer some degree of shelter for free exchange. Put differently, had BioGovernance Commons been framed as a civic forum rather than an academic one, many China-based participants would likely have been reluctant to take part. It is precisely under the premise of 'academic discussion' that participants feel both permitted and obligated to speak openly. The irony, of course, is that this careful demarcation between 'academic' and other forms of debate emerges precisely because academia itself is deeply imbricated with political intervention. The space of academic exchange is thus simultaneously constrained and protected: it enables a degree of openness not despite political oversight, but because it is institutionally recognised as a legitimate and carefully circumscribed mode of discourse.

Within this context, writing that departs from formal academic conventions poses a particular challenge for some Chinese contributors. Such departures blur boundaries between the personal and the formal, between scholarly work and public commentary. Contributors understandably worried about how their writing might be perceived by institutions and authorities: would they still be seen as speaking *as academics*? Or speaking out of turn? At the same time, it is precisely academic formality, largely shaped by Western epistemic norms, that often produces epistemic injustice by marginalising alternative modes of expression, reflection, and knowledge-making. Its reductive tendencies leave little room for other ways of writing and thinking, limitations this book explicitly seeks to resist.

In this sense, an unexpected challenge for the editors was not simply logistical delay, but the difficulty some contributors faced in 'writing otherwise'. That difficulty, however, is itself analytically generative. It reveals how deeply questions of form, voice, legitimacy, and risk are embedded in contemporary knowledge production and why commoning practices, fragile as they are, remain indispensable.

It is also in this context that we are particularly grateful for the support and commitment of Professor Ino Rossi and the editorial team at Springer in supporting this volume. What the book captures are necessarily partial snapshots of the conversations, tensions, and reflections that emerged through the BioGovernance Commons—glimpses of what we attempted and, in some instances, what we managed to do well. At the same time, there are ideas, exchanges, and moments of insight that are inevitably absent from these pages, yet remain vibrant and generative within the lived events of the Commons itself.

This recognition points to a broader reflection: the book, or writing more generally, is neither the sole nor necessarily the most adequate medium through which such experiments should be documented or extended. Rather, we hope readers will approach this volume as an invitation, perhaps especially because of its imperfections, to imagine and experiment with other ways of rendering visible different forms of collective inquiry that seek to move beyond entrenched geopolitical and epistemic boundaries. It is for this reason that we believe additional contextualisation may help readers better appreciate not only what is included in this book, but also what necessarily exceeds it.

Structure of the Book and Additional Context

The volume appropriately opens with Lu (Lucy) Gao's chapter on 'humanising' China's COVID-19 research (Chap. 2). This contribution offers fresh insights into one of the core concerns that the BioGovernance Commons initiative seeks to address—namely, how to make sense of China's life sciences as they are actually practised. It also exemplifies the generative impact of the initiative's collective discussions, demonstrating how sustained dialogue can inspire new research questions and motivate original empirical inquiry.

Trained in philosophy and specialising in the history of science, Gao directs the Science, Technology and Society Research Centre at the Institute for the History of Natural Sciences, Chinese Academy of Sciences. She is deeply experienced in science governance and emerging technology policy, serving in key leadership roles within national academic associations, and has held visiting positions at Stanford University in the United States and at the Universities of Edinburgh and Kent in the United Kingdom. As a core member of the BioGovernance Commons initiative, she has contributed to nearly every gathering, bringing both intellectual depth and extensive policy experience to the collective.

As Gao notes at the outset of the chapter, the widening global divide—accelerated by asymmetrical media portrayals of the pandemic both within and beyond China—was a key impetus for the monthly BioGovernance Commons dialogues. Her research can be seen as one of the most tangible outcomes of these exchanges. Through collective discussion, participants came to recognise that the problem was not limited to one-sided representations in Western media. Chinese scientists' perspectives and lived experiences during the pandemic were also marginalised within

China's domestic media landscape, where safer, prescribed nationalist narratives tended to prevail. In prioritising the 'national image', the human dimensions of scientific and social responses to the pandemic were paradoxically—and in some cases deliberately—obscured.

Crucially, the BioGovernance Commons provided a safe space for open questioning and explanation among global peers. This enabled participants such as Gao not only to reassess the presumed unbridgeability of divides between China and the West, but also to recognise that part of the difficulty lies in the limitations of existing normalised thinking 'habits' for categorising and interpreting difference. As Gao reflects,

I often ask myself why I prefer discussing ethical theories and issues in the policy domain to telling stories, or why I am more willing to focus on global issues with universal significance while reducing China-specific issues to the conclusion that "they are special in China", or attributing them solely to the country's stage of scientific, technological, and social development (Gao, Chap. 2).

It is against this backdrop that Gao undertook interviews with Chinese scientists directly involved in COVID-19 research, seeking to document their experiences, frustrations, and aspirations—many of whom also felt misrepresented or silenced within domestic media narratives. While the chapter gives voice to these individuals, it also goes further. As the chapter title, 'Deconstructing Nationalist Science: Humanising COVID-19 Research in the Era of the Whole-Nation System', indicates, the piece offers a critical interrogation of how competing and conflicting stories emerge within China's contemporary science governance framework and what it might take to meaningfully understand them.

Foregrounding human experience in critical inquiry is foundational to any practice of commoning. For this reason, Gao's empirically grounded and reflexive contribution is a fitting chapter for the volume. It is also worth noting that her call to 'humanise science' did not end with this study. Between 2024 and 2025, she further developed these ideas by co-editing a special collection entitled 'Humanising Responsible Research and Innovation in a Cosmopolitan World: A China Perspective' for the *Journal of Responsible Innovation*, together with Ping Yan (a co-host of the 2025 BioGovernance Commons meeting in Dalian) and Tom Xiaowei Wang. This extended engagement underscores both the intellectual durability and the broader relevance of the themes introduced in this chapter.

Such humanising effort, as Gao's chapter and her other publications (see Gao forthcoming) have shown, is not limited to efforts required from the West, but it also echoes a point that China needs to 'render "epistemic justice" to its own people' (Centre for Global Science and Epistemic Justice 2022, 12).

Bringing 'the human story' back into academic and socio-political discourse *on* and *of* a 'global China', a country with which almost everyone has 'inherent and manifold interconnections' (Chap. 3), is also the central aim of Daniela Caterina's contribution, '*Afraid of whom? A European outlook on China between research priorities and biographical encounters with the Other*'.

At the time of the BioGovernance Commons discussions, Caterina was an Associate Professor in the School of Philosophy at Huazhong University of Science and Technology in Wuhan, and one of the few European scholars based in China. Despite the chapter's subtitle referring to a 'European outlook', it would be both difficult and misleading to categorise her position simply as that of an Italian scholar offering either an insider–outsider or outsider–insider perspective. Caterina is, instead, a genuinely cosmopolitan academic, whose research spans linguistic, state theory, political economy, critical political economy, and discourse analysis—and whose personal and professional life was shaped by COVID-19 lockdowns experienced in both China and Europe.

What makes this chapter particularly compelling, and at times unsettling, is its refusal to address the long-standing Western preoccupation with whether we should be 'afraid of China' (Barr 2011; Pavlićević and Talmacs 2022) at the level of abstract geopolitics alone. Instead, Caterina approaches this question through the lens of her own biographical trajectory as an academic migrant to China, intertwined with her participation in the BioGovernance Commons. In doing so, she seeks 'to zoom in on what China really means for the people who experience it in their everyday life' (Chap. 3).

Despite, or perhaps because of, its deeply personal tone, the chapter extends Caterina's earlier Gramscian analyses of China (see Caterina 2021, 2025). Drawing on the tightening of surveillance and regulatory scrutiny over European research collaborations with China, she highlights in the chapter how debates about research ethics, autonomy, and responsibility are increasingly shaped by broader geopolitical fragmentation. The chapter poses a series of incisive questions: 'What is the "China" to which we refer, and with what implications?' Is China's science understood as a 'model' to be examined, avoided, emulated, or resisted? Or has 'China' instead become a rhetorical device, an 'ideal enemy', mobilised to generate cohesion among otherwise fragmented political constituencies?

Crucially, Caterina moves beyond a stance of detached observation. Reflecting on her lived experience, she writes: 'I was not simply living in China anymore; I was living *China* and allowing the Chinese Other—through its very presence—to leave new traces in my vision of the world.' This encounter with the Other, she suggests, is always more generative and transformative than we anticipate.

Caterina is not uncritical of China and its hegemonic logic, but for her, the 'China question' ultimately *interrogates* Europeans themselves—asking whether the internal contradictions of modern capitalism, science, and politics can truly be resolved by 'solving China'. At the same time, it *concerns* Europe because China, given its scale of production and consumption, is inseparable from planetary human and non-human wellbeing. Most poignantly, she concludes that the China question 'challenges us as human beings, reminding us of the need to reach out to the people and restore the centrality of dialogue and one-to-one exchange.'

However, it is important to bear in mind that the 'China Question' itself has evolved. Arguably, it has shifted from Joseph Needham's famous fascination with why modern science did not emerge in China to a contemporary concern framed increasingly around China as a potential threat. Yet, even two decades ago, at the turn of the millennium—when China was only beginning to emerge as a scientific powerhouse—the China Question took a rather different form. It centred on China's potential and on whether the country possessed the 'right culture' to harness that potential in order to become a valued participant in the 'cosmopolitanisation of science' (see Salter et al. 2006; Sleeboom-Faulkner 2010; Zhang 2012). The questions, critiques, and effects that emerged through transnational dialogue at that time—along with the underlying power dynamics shaping how China was known—were markedly different from those that prevail today. It is in this context that we are particularly fortunate to include Haidan Chen's contribution in Chap. 4, which offers an overview of this dialogical shift over the past two decades, examined primarily through debates surrounding biobanks.

Indeed, one topic that has retained particular relevance in discussions of science governance in and with China is biodata. This theme has also been central to BioGovernance Commons discussions. Not least, this is because processes of 'commoning' are critical for the interoperability of biodata, which in turn is essential for its assetisation. In fact, prior to COVID-19, Chinese academic communities had already made efforts to 'common', if not standardise, data collection and codification through initiatives such as the Belt and Road Life Sciences Economy Alliance (www.brlsea.org), seeking to assert and sustain China's comparative advantage and leadership in this field. At the Alliance's first General Assembly in December 2020, the establishment of a Belt and Road Regional International Standardisation Committee was announced alongside the declaration that 'a new round of globalisation' had begun (LSEA 2020). This signalled a more overt phase of South–South collaboration led by Chinese authorities and scientific communities, one that directly challenges prevailing norms concerning how standards are defined and who is entitled to shape them (Zhang and Datta Burton 2022).

A further, and perhaps more pressing, reason for biodata's prominence lies in the COVID-19 pandemic itself, during which data sharing became both essential and politically contentious. This was also a period in which China—much like several Global South countries such as Mexico, Brazil, and India (Benjamin 2009; Kent et al. 2015)—simultaneously accelerated its biobanking efforts while tightening regulatory control over human genetic materials (Zhang 2023). These developments can be understood both as an overdue response to historical harms associated with colonial biopiracy and as an immediate reaction to perceived global antagonism and political hostility towards Chinese science. As a consequence, some of the initially proposed regulations governing reporting systems and benefit-sharing protocols proved excessively strict and impractical, rendering international collaboration extremely difficult (Chen and Li 2025). Members of the BioGovernance Commons group—including Ruipeng Lei, Di Zhang, Leifan Wang, Joy Zhang, and others—contributed to a collective set of 'Policy Recommendations for China's Implementing Rules for Human Genetic Resources Management', submitted to China's Ministry of Science and Technology. The core argument advanced was the need for more evidence-based and practice-oriented regulation that would be both enforceable and more conducive to public health and the development of Chinese science (see Lei et al. 2022).

Few scholars are better positioned to analyse the co-evolution of international dialogue and biobanking in China than Haidan Chen, Associate Professor with tenure at the Department of Medical Ethics and Law, School of Health Humanities, Peking University. Chen received her PhD in Philosophy of Science and Technology from Zhejiang University and has been a visiting scholar at the University of Edinburgh, the Brocher Foundation in Switzerland, and the Center for Biomedical Ethics at Stanford University as a Fulbright scholar.

Chen not only authored one of the earliest academic analyses of biobanking in China (Chen 2011) but also played an active role in BIONET, one of the earliest EU–China collaborations on BioGovernance, led by Nikolas Rose during his tenure as Director of the BIOS Centre at the London School of Economics and Political Science. BIONET represented one of the first transnational initiatives to promote constructive and mutually reflexive dialogue on life-science governance with China, and it stands as a key precursor to the BioGovernance Commons model.

To some extent, biobanking is particularly vulnerable to reductive rhetoric. It is easily framed within nationalist discourse as a capitalist, extractive, and productivity-driven enterprise. Yet, at the same time, it resists such simplification. As Chen notes in Chap. 4, 'the life cycle of biobanks involves numerous individuals, agencies, organisations, and institutions in the collection, storage, distribution, and use of biological samples and associated data'. It is therefore essential to examine both the Chinese authorities' top–down regulatory approach (i.e. treating biobanks as human genetic resources and strengthening protection in the name of national security) and the dense social networks and active processes of social learning developed by individual biobanks.

Chen reflects candidly: 'My fieldwork further reveals a significant gap between the formal regulatory framework, which emphasises state control, legal codification, and the securitisation of genetic resources against perceived threats, and the practical enactment of these regulations by researchers managing biobanks, which is primarily driven by immediate research needs, collaboration, and sustainability. My long-term involvement in this field—from the early experiences of BIONET to more recent engagement with the BioGovernance Commons—underscores these tensions. Both initiatives provided platforms for cross-cultural dialogue, critical reflection, and mutual learning, creating experimental spaces for governance innovation… However, these dialogical and experimental approaches remain peripheral to China's governance landscape.' For China to realise its ambitions in the life sciences—for the benefit of both China and the wider world—Chen argues that a cultural shift is required: one that recognises the legitimacy of bottom-up initiatives, patient and public voices, and sustained international collaboration.

As founders of the BioGovernance Commons, we hope and believe, that *commoning* may offer one pathway toward initiating such a cultural shift. However, for commoning to be meaningful and durable, it must be pursued in concert with other socio-political practices and institutional mechanisms. This includes careful attention to differences in legal systems, a concern that comes to the fore in Chaps. 5 and 6.

Continuing the discussion on data governance, legal expert Li Du (Macau University) and his colleague Zhangyu Wang offer a thorough and well-documented account of the 'multi-layered, complex, and at times contradictory' regulatory regime surrounding the cross-border transfer of human genetic data (HGD) in China, situating governing challenges within both national security imperatives and global scientific collaboration (Chap. 5). Li Du is an Associate Professor at the University of Macau Faculty of Law and one of the early participants of BioGovernance Commons. Trained in both China and Canada, Du has led many research projects on legal and ethical implications of novel and emerging technologies, e.g., innovative biotechnologies, e-health, and generative AI in biomedical research.

Du and Wang's contribution highlights that, while China has made significant efforts to clarify its regulatory framework for human genetic data into a tripartite structure, governance in practice remains fragmented. The enforcement of HGD regulations relies on multiple domains—including medical regulation, cybersecurity law, data protection, economic planning, ethics review mechanisms, and national security—whose respective authorities do not always coordinate effectively. Legislative incoherence persists, alongside overlapping obligations and varied, often ambiguously defined, assessment requirements. As a result, regulatory fragmentation and internal tension are not merely transnational challenges but also deeply domestic ones: different legislative and administrative frameworks within a single nation-state must, yet frequently fail to, 'speak' to one another. From a commoning perspective, this raises a more fundamental governance question—not simply how data are controlled or transferred, but how regulatory regimes can be rendered mutually intelligible and relationally aligned.

For Du, China's HGD regime remains largely defensive and inward-looking. China should continue to align with international movements toward research data sharing and, as an emerging power, take a leading role in articulating and advancing paradigms for open, responsible, and globally engaged scientific data governance. To 'lead', however, is not simply about developing and promulgating new rules from a Chinese point of view, but should rely on inclusive and iterative dialogues.

It is also worth noting that Du and his team founded *iBioLaw*, a weekly WeChat bulletin that translates and curates global policy, legal, and ethical developments in the life sciences for Chinese audiences, serving as an important knowledge-bridging platform.

Chapter 6, 'China's Role in Global Risk Governance', is written by Simona Novaretti, Associate Professor of Comparative Private Law at the University of Turin, School of Law, where she teaches Chinese Law, Comparative Law, and Law and Society in Asia. Novaretti is also the author of *Law and Bioethics: The Beginning of Life in China* (2025). Novaretti has contributed to both post-COVID in-person BioGovernance Commons meetings in China: the 2023 Chengdu meeting on 'Bioeconomy for the Common Good', organised by Ruipeng Lei, and the 2025 Dalian meeting on 'Chinese Modernisation and Global Governance', organised by Ping Yan. Her sustained engagement with the initiative places her chapter firmly within the dialogical ethos of the Commons.

Her work addresses a significant knowledge gap in understanding the relationship between law and bioethics in China. Through an examination of ethical principles, legal norms, and judicial practice relating to biotechnologies involved in human reproduction and the beginning of life, Novaretti demonstrates how transnational dialogue and cooperation on shared ethical challenges do not undermine Western legal thinking. On the contrary, engagement with non-Western legal traditions can enrich and expand existing normative frameworks.

Her chapter offers fresh insights into the so-called 'Wild East' problem that China has been grappling with since the turn of the millennium, namely, the enduring characterisation of China as a country 'without': without God, or without law. Novaretti directly confronts this reductive framing and examines how China, through its evolving legislative framework and regulatory culture, may contribute meaningfully to the global commons for the common good, adopting a decolonial perspective on BioGovernance. For her, the promise lies in Articles 109 and 990 of the PRC Civil Code, enacted during the COVID-19 pandemic in the aftermath of the CRISPR baby scandal and effective from 1 January 2021:

> **Article 109**
> The personal freedom and personal dignity of natural persons are protected by law.
> **Article 990**
> Personality rights are the rights enjoyed by civil subjects, including the right to life, bodily integrity, health, name, designation, portrait, reputation, honour, and privacy. In addition to the personality rights listed above, natural persons enjoy other personality interests arising from personal freedom and personal dignity.

Novaretti argues that the incorporation of personal freedom and personality rights into the highest level of civil norms enables judges to rely on these provisions directly in adjudication. This is already evident in judicial decisions concerning clinical practices such as posthumous reproduction and gestational surrogacy. Moreover, she highlights how the Chinese legal understanding of the individual—as inherently relational and embedded within family, society, the state, and humanity as a whole—offers conceptual resources that may be productively mobilised at the global level. Such a perspective can help build a BioGovernance system capable of mediating across diverse legal and cultural traditions, rather than defaulting to Western normative assumptions.

Yet, like many participants in the BioGovernance Commons, Novaretti is not naïve about the challenges involved. She cautions that open, transparent, and epistemically inclusive dialogue remains essential. More concretely, she writes that 'China must take certain steps, starting with small things'. Reflecting on her own experience as a scholar of Chinese law based outside China, she notes that researching China has become increasingly difficult in recent years. The quiet removal of seemingly minor details from the public domain, she warns, reveals that prejudice and mistrust are not exclusively Western phenomena.

'Value difference' and 'value conflict' may appear to be two sides of the same coin. However, one of the first points clarified in the BioGovernance Commons meetings was that discussions of 'values' often concern *prioritisation strategies under conditions of resource constraint*. While all societies face limited resources, an obvious but frequently overlooked fact is that societies at different socio-economic stages and on different development trajectories experience *different forms* of resource limitation. One of the implicit questions one may have in mind when thinking about global governance is whether China remains a 'developing' or 'Global South' country. This is certainly one main contention in all of BioGovernance Commons discussions. To be sure, positioned simultaneously as a rising global scientific power and an upper-middle-income developing country (UNCTAD 2025; World Bank 2025), China occupies a liminal space between Western scientific centres and the wider Global South, while still contending with the enduring legacies of exploitative medical and bioprospecting practices (Keim 2003; Zhang and Datta Burton 2022).

Epistemically, in terms of whose knowledge is recognised as valid and agenda-setting, however, China continues to occupy a structurally disadvantaged position within global knowledge hierarchies. Many Chinese academics are caught between the aspiration to articulate new pathways of governance and contribute original interventions to global debates, and the necessity of navigating a dual set of constraints: on the one hand, the continued reliance on Western theoretical frameworks as a relatively secure route to academic credibility and peer recognition; on the other, the boundaries of domestic political tolerance that delimit what is considered legitimate and permissible scholarly space.

The question of how to move beyond this double constricture was a central motivation for Dalian University of Technology, one of China's leading national hubs for research in science and technology ethics, science and technology management, and science and technology policy, to convene an aforementioned further in-person BioGovernance *Commons* meeting in late August 2025.

The symposium foregrounded the urgency of articulating China's indigenous pathways to governance and modernity. Zhili Liu, Dean of the School of Marxism at Dalian University of Technology and host of the meeting, together with a number of keynote speakers—many of whom serve as deans or heads of schools at leading social science institutions across China—emphasised China's aspirations to contribute to a more just and equitable international order across multiple domains. Collectively, they called on scholars to engage in open and inclusive dialogue, and to advance rigorous, evidence-based proposals capable of providing both theoretical foundations and practical pathways for integrating Chinese-style modernisation with global governance. For example, Jiuyu Xiang, Director of the Institute for the Study of Mainstream Ideology at Wuhan University, argued that common prosperity should be elevated to a core ethical objective of modernisation, serving as a normative anchor for reconfiguring global development patterns and conceptions of justice. In this account, Chinese-style modernisation is a broader strand of contemporary Chinese policy thought in which modernisation is understood not simply as economic growth, but as a normatively guided process oriented towards 'common prosperity'—that is, the coordinated pursuit of development, social equity, and collective well-being. In this framing, Chinese-style modernisation is a state-led and ethically grounded project that seeks to recalibrate the developmental model while also positioning itself as a civilisational alternative to Western trajectories. It is neither a passive response to nor a simple replacement of Western models; rather, it constitutes a reassertion of civilisational agency—entailing a critical reworking of modernity itself and offering a paradigmatic innovation of historical significance for global governance.

Yet, despite the normative ambition of this vision, what such a pathway might concretely entail remains largely underspecified. More importantly, no discussion on building a BioGovernance commons with China would be appropriate without considering China's regional impact. Thus, in the most recent Dalian meeting, in addition to contributors from Europe and North America, we made a point in inviting scholars from China's neighbouring countries, such as Japan, India, and Malaysia. This helps to further break the West vs China binary and produce more nuanced consideration of what it means to be common, what needs to be 'commoned' for a coordinated governance to function efficiently, and just as important, commoning with whom, how to identify and cultivate stakeholders.

We revisited some of the key themes of BioGovernance Commons discussions, but with fresh regional insights from non-Western and non-Chinese perspectives. We're pleased to include two such contributions in this book.

Chapter 7 analyses biosecurity governance. Biosecurity emerged as a central theme very early in the BioGovernance Commons discussions. This was partly due to timing: the initiative began during the COVID-19 pandemic, when lab-leak theories were widespread, and forms of research such as gain-of-function studies had become the subject of intense regulatory, political, and public concern.

Equally important, however, was a significant policy development that occurred shortly after the BioGovernance Commons was launched. In July 2021, China introduced the *Tianjin Biosecurity Guidelines for Codes of Conduct for Scientists* (Center for Biosafety Research and Strategy et al. 2021). Notably, this initiative was championed not by China's Ministry of Science and Technology, but primarily by the Ministry of Foreign Affairs—signalling biosecurity as a matter of international governance and diplomatic engagement, rather than solely domestic scientific regulation.

We were particularly fortunate to engage directly with Leifan Wang and colleagues from Tianjin University, who played a critical role in developing the Guidelines. Their participation enabled an in-depth discussion of the rationale underpinning the document, especially the framing of human rights as a potential shared normative foundation for global biosecurity governance.

Two insights from these discussions are especially noteworthy. First, as documented elsewhere (Zhang et al. 2023a), it was Chinese participants who voiced the most critical reflections. Their concerns centred on whether—and how—the Guidelines might meaningfully shape everyday laboratory practice, particularly in the context of China's expanding academia–industry collaborations and regulatory incentives for societal investment in translational research.

Second, questions of enforceability loomed large. Participants highlighted the uneven applicability of such guidelines across diverse regulatory environments, especially amid shifting power dynamics driven by the growing influence of the bio-industry and venture capital. These asymmetries, alongside wide variation in national regulatory capacity, complicate efforts to balance the safe development of science with the promotion of locally led biomedical research—especially where biosafety and biosecurity risks are entangled with economic and innovation agendas.

It was in this context that we invited Suryesh Kumar Namdeo to broaden the discussion on how a BioGovernance commons might illuminate new pathways for transnational coordination on biosecurity. Namdeo is a Senior Research Analyst at the Indian Institute of Science, Bengaluru. Although not a China specialist, he previously served as a science and technology consultant to the Biological Weapons Convention at the United Nations Office for Disarmament Affairs in Geneva, and specialises in science diplomacy, biosecurity, and open science. His chapter goes beyond a West–East binary, offering nuanced reflections drawn from experiences across China, India, Nigeria, and Singapore. In doing so, it foregrounds the uneven, relational, and context-sensitive nature of biosecurity governance, while demonstrating how a commons-oriented perspective can help reframe coordination challenges that are often obscured by conventional geopolitical or regulatory framings.

Last but not least, Science Communication and Knowledge Networks specialists Hong Ngen Fung and Mofei Lyu's analysis of the bilateral BioGovernance Commons of Traditional Chinese Medicine between China and Malaysia offers a critical reminder that *commoning* is not confined to emerging or cutting-edge biotechnologies, but is equally necessary for the governance of more established sciences and technologies (Chap. 8). This is because commoning should not be understood merely as a spatial practice for accommodating identifiable value conflicts, but rather as a fundamentally temporal process. The chapter demonstrates that strategic policy rhythms and proactive synchronisation are central to the making of a BioGovernance Commons, operating across four interrelated axes of governance: epistemic legitimation, regulatory alignment, knowledge transmission, and discursive amplification.

Another point to highlight from Fung and Lyu's chapter is its insistence that good governance must embed the planning of effective public communication, rather than treating it as a parallel or downstream activity. This is critical for 'enhancing public awareness and societal confidence' (Chap. 8) in ways that address individual uptake barriers and, in turn, improve public health.

Removing uptake barriers, however, is fundamentally different from promoting—or worse, enforcing—a particular technological pathway. Far from dulling critical edge, enhancing awareness and confidence is about expanding it: giving individuals and communities meaningful access to information, deliberative space, and the capacity to discern and choose among a full range of technoscientific solutions. In this sense, public communication is not a tool of persuasion but a condition of agency. At the time of writing, Fung is organising an international conference on *Science, Technology and Innovation for Social Good: Adapting Policy, Society and Industries for Sustainable Development*, to be held in July 2026 in Kota Kinabalu, Malaysia. As an extension of the BioGovernance Commons meeting in Dalian, Fung has developed a dedicated stream on *Global Public Deliberation in Science and Technology: Knowledge Pluralism and the Policy Interface*. This stream is conceived as a space that brings together academics, policymakers, and thought leaders to critique, extend, deepen, and reimagine practices of global public deliberation—particularly in relation to Asia and the Global South.

The final chapter (Chap. 9) reflects on two developments that emerged during the closing stages of this volume: limitations placed on Chinese involvement in Horizon Europe and an instance in which unequal access to research databases led to divergent scholarly conclusions. Drawing on these examples, the chapter contends that knowledge production today is increasingly shaped by geopolitical divisions and uneven research infrastructures. Such conditions not only hinder collaboration but also influence how intentions, credibility, and trust are perceived across different settings.

Zhang characterises the future of the BioGovernance Commons as 'pessoptimistic'—a condition defined simultaneously by the weakening of shared epistemic spaces and the continued need to sustain them. From this perspective, 'commoning' is understood as an ongoing, practice-oriented effort to maintain dialogue amid difference, requiring continuous attention and relational labour to uphold the fragile conditions that make collaborative understanding possible.

It has been challenging to include natural scientists in this volume, not least because of the substantial social science orientation and writing requirements of this book series. Nevertheless, we are delighted to feature contributions from two scientists who have both served as reviewers and have each provided a short commentary reflecting on what the BioGovernance Commons initiative—and similar efforts—could and should become.

These two contributors represent different career stages and bring complementary scientific and governance experience. Dr. Wei He, who completed his doctoral training in the UK, is now based in Xi'an, a major educational and innovation hub in north-west China. While his primary research lies in plant science, in recent years, he has devoted significant effort to advancing international student exchange, postgraduate training, and research collaboration between his institution and global partners. Although he did not participate directly in BioGovernance Commons activities, he played an important role in shaping the earlier UK–China dialogues on public engagement in the life sciences (2017–2019; see Zhang 2018), which helped to establish the foundational networks and trust underpinning the initiative. His contribution (Coda I) offers a valuable 'outside insider' perspective.

The second commentary is provided by Professor Xian-En Zhang, a highly respected synthetic biologist in China who previously served as Director-General for Basic Research at the Ministry of Science and Technology. He has been directly involved in a couple of key in-person BioGovernance Commons events. Drawing on his experience across both research and science governance, his contribution (Coda II) offers an informed perspective on how transnational initiatives of this kind may be viewed at senior levels of the scientific community. It reflects on possible future directions for the BioGovernance Commons and similar efforts.

As Françoise Baylis, President of the Royal Society of Canada, aptly observed, 'there's only one real question, and that question is what kind of world do we want to live in? Once we've answered that, then it's time to look at all the [science] and say, how will this help me build that world?' (McCabe 2025). We see the

BioGovernance Commons as an experiment, a developing platform designed to support individuals and collectives in grappling with precisely this question, and, just as importantly, as a dialogical mechanism for fostering coordination among diverse and sometimes competing answers. We neither had the capacity nor the need to document every conversation that took place over the past few years. However, we did produce reports on two major ethically and politically contentious issues in human genome editing—conversations that would not have been possible to convene without the trust relationships built through the BioGovernance Commons. These two global discussions, one in 2022 and the other in 2023, were not intended to produce consensus, but rather to enable transparent and candid exchange between Chinese and international peers. Both initiatives emerged from perceived divergences in approaches between China and its global counterparts, and the reports deliberately preserve the disagreements that remained unresolved. At the same time, these exchanges helped to foster a degree of coordination and mutual understanding. We include these reports as appendices to this volume in the hope that they offer readers a glimpse of the BioGovernance Commons in action.

We are deeply grateful to those who have contributed to making this a productive endeavour, who have helped extend it to wider audiences, and who have expanded our own horizons in the process. Yet, we do not believe there is a single correct way of doing this work. Instead, we hope the discussions collected in this volume will spark new ideas and practices for understanding and engaging with one another, including with actors and institutions often perceived as opaque or radically different from ourselves.

References

Barr, M. T. 2011. *Who's afraid of China?: The challenge of Chinese soft power*. London: Zed Books.

Ben Ouagrham-Gormley, S., and K. M. Vogel. 2020a. Scientific advocacy and dissent in China and Russia: Role of scientists in shaping communication and governance around human germline genome editing. *Health Security* 18 (6): 444–457.

Ben Ouagrham-Gormley, S., and K. M. Vogel. 2020b. Follow the money: What the sources of Jiankui He's funding reveal about what Beijing authorities knew about illegal CRISPR babies, and when they knew it. *Bulletin of the Atomic Scientists* 76 (4): 192–199. https://doi.org/10.1080/00963402.2020.1780726.

Benjamin, R. 2009. A lab of their own: Genomic sovereignty as postcolonial science policy. *Policy and Society* 28 (4): 341–355. https://doi.org/10.1016/j.polsoc.2009.09.007.

Bharadwaj, A. 2012. Enculturating cells: The anthropology, substance, and science of stem cells. *Annual Review of Anthropology* 41:303–317.

Caterina, D. 2021. Gramsci in China: Past, present, and future of a still open encounter. *Antipode* 53 (5): 1357–1376.

Caterina, D. 2025. Clashing narratives over Italy's adhesion to the belt and road initiative: A critical discourse analysis approach. In *Narrating China and Europe in uncertain times. 21st century standpoints*, edited by Á. Révész, D. Freeman, M. Feldmann, and S. Langendonk, 36–56. Bristol: Bristol University Press.

Center for Biosafety Research and Strategy, Johns Hopkins Center for Health Security and The InterAcademy Partnership. 2021. *Tianjin biosecurity guidelines for codes of conduct for scientists*. Available at: https://www.interacademies.org/sites/default/files/2021-07/Tianjin-Biosecurity-Guidelines-Codes-Conduct.pdf

Centre for Global Science and Epistemic Justice. 2022. *The ethics of protecting 'CRISPR babies': An international roundtable summary report*. Canterbury: University of Kent.

Chen, H. 2011. Cord blood banking in China: Public and private tensions. *East Asian Science, Technology and Society: An International Journal* 5:329–339.

Chen, J., and W. Li. 2025. Is cross-border transfer of China's human genetic data an impossible mission? *Human Genetics* 144 (7): 715–725. https://doi.org/10.1007/s00439-025-02756-8.

Cyranoski, D. 2018. CRISPR-baby scientist fails to satisfy critics. *Nature* 564:13–14. https://doi.org/10.1038/d41586-018-07573-w.

Frontiers of Bioethics Law. 2023. *Consensus statement of the workshop on challenges and responses to ethics governance in science and technology*. WeChat Official Account. Available at: https://mp.weixin.qq.com/s/RN3mElJqzYbFoIbJIJ72CQ

Gao, F. forthcoming. Humanising RRI in China. *Journal of Responsible Innovation*.

Hamilton, Jennifer A., Banu Subramaniam, and Angela Willey. 2017. What Indians and Indians can teach us about colonisation: Feminist science and technology studies, epistemological imperialism and the politics of difference. *Feminist Studies* 43:612–623.

Kelly, É. (2021) Talks on future science partnership with China "not an easy exercise". *Science|Business*. Available at: https://sciencebusiness.net/technology-strategy-board/news/talks-future-science-partnership-china-not-easy-exercise

Keim, B. 2003. Out of sight, out of mind: How Harvard University exploited rural Chinese villagers for their DNA. *GeneWatch* 16:10–11.

Kent, M., et al. 2015. Building the genomic nation: "Homo Brasilis" and the "Genoma Mexicano". *Social Studies of Science* 45 (6): 839–861. https://doi.org/10.1177/0306312715611262.

Lei, R., H. Jiao, H. Liu, J. Shi, J. Y. Zhang, H. Zhang, P. Yao, C. Wu, D. Zhang, Y. Wang, et al. 2022. *Policy recommendations for China's "implementing rules for human genetic resources management"*. Wuhan: HUST Center for Bioethics.

Lewis, M. K. 2020. Criminalizing China. *Journal of Criminal Law and Criminology* 111:145–202.

Life Sciences Economy Alliance (LSEA). 2020. *The second session of the 1st belt and road life sciences economy Alliance general assembly was successfully held at CNGB, Shenzhen*. Available at: http://www.brlsea.org/article/24 (Accessed: date).

Liu, Z., J. Shi, and J. Xu. 2024. He Jiankui's unprecedented offense and worrying comeback: How the CRISPR-babies scandal reshaped the legal governance of scientific research in China. *Journal of Responsible The Innovation* 11 (1): https://doi.org/10.1080/23299460.2024.2372116.

McCabe, D. 2025. *Encouraging scholars to step into the spotlight*. McGill News. Available at: https://mcgillnews.mcgill.ca/encouraging-scholars-to-step-into-the-spotlight/.

Mitchell, S. D. 2009. *Unsimple truths: Science, complexity, and policy*. Chicago, IL: University of Chicago Press.

Nightingale, A. J. 2011. Beyond design principles: Subjectivity, emotion, and the (ir) rational commons. *Society & Natural Resources* 24 (2): 119–132. https://doi.org/10.1080/08941920903278160.

Ostrom, E. 1990. *Governing the commons: The evolution of institutions for collective action*. Cambridge: Cambridge University Press.

Pavlićević, D., and N. Talmacs. 2022. Answering the "China question": Local responses to global China. In *The China question: Contestations and adaptations*, edited by D. Pavlićević and N. Talmacs, 1–20. Springer Nature: Singapore.

Roy, A. 2016. Who's afraid of postcolonial theory. *International Journal of Urban and Regional Research* 40 (1): 200–209. https://doi.org/10.1111/1468-2427.12274.

Salter, B. 2008. Governing stem cell science in China and India: Emerging economies and the global politics of innovation. *New Genetics and Society* 27 (2): 145–159.

Salter, B., M. Cooper, and A. Dickins. 2006. China and the global stem cell bioeconomy: An emerging political strategy? *Regenerative Medicine* 1 (5): 671–683.

Sleeboom-Faulkner, M. E. 2010. National risk signatures and human embryonic stem cell research in mainland China. *Health, Risk & Society* 12 (5): 491–511.

Singleton, B. 2017. What's missing from Ostrom? Combining design principles with the theory of sociocultural viability. *Environmental Politics* 26 (6): 994–1014. https://doi.org/10.1080/09644016.2017.1364150.

Trimble, M. 2017. US leads world in scientific research output, but dominance shrinking: Study. *U.S. News—Best Countries*. Available at: https://www.usnews.com/news/best-countries/articles/2017-06-16/us-leads-world-in-scientific-research-output-but-dominance-shrinking-study

UNCTAD. 2025. *Countries, all groups hierarchy*.

Vogel, K. M., and S. Ben Ouagrham-Gormley. 2022. Scientists as spies? Assessing U.S. claims about the security threat posed by China's thousand talents program for the U.S. life sciences. *Politics and the Life Sciences* 41 (1): 1–33. Available at: https://www.cambridge.org/core/journals/politics-and-the-life-sciences/article/abs/scientists-as-spies/DA5B1DD06F939076B5CD0D72423FB025.

World Bank. 2025. *Country overview: China*. Available at: https://www.worldbank.org/en/country/china/overview

Zhang, J. Y. 2012. *The Cosmopolitanization of science: Stem cell governance in China*. Basingstoke: Palgrave Macmillan.

Zhang, J. Y. 2015. The "credibility paradox" in China's science communication: Views from scientific practitioners. *Public Understanding of Science* 24 (8): 913–927.

Zhang, J. Y. 2018. *Governing scientific accountability in China. Final report of the ESRC research project*. Canterbury: GSA-China. https://kar.kent.ac.uk/77924/1/GSA-Project-Report.pdf.

Zhang, J. Y. 2023. What do the British and Chinese governing visions on human genomic research tell us about biosovereignty? *Asian Biotechnology & Development Review* 25 (1–2): 5–23.

Zhang, J. Y., and M. Barr. 2019. The transformative power of commoning and alternative food networks. *Environmental Politics* 28 (4): 771–789. https://doi.org/10.1080/09644016.2018.1513210.

Zhang, J. Y. 2021a. To keep nationalism in check, nurture science solidarity. *Nature* 591:9.

Zhang, J. Y., and M. Barr. 2021. Harmoniously denied: COVID-19 and the latent effects of censorship. *Surveillance & Society* 19 (3): 389–402.

Zhang, J. Y., and S. Datta Burton. 2022. *The elephant and the dragon in contemporary life sciences: A call for Decolonising global governance*. Manchester: Manchester University Press.

Zhang, J. Y., S. Ben Ouagrham-Gormley, and K. Vogel. 2022. Creating common ground with Chinese researchers. *Issues in Science and Technology*45–48.

Zhang, J. Y., S. Ben Ouagrham-Gormley, and K. Vogel. 2023a. Chinese academics are becoming a force for good governance. *Issues in Science and Technology* 39 (4): 81–84.

Zhang, J. Y, Ruipeng Lei, Kathleen Vogel, Sonia Ben Ouagrham-Gormley 2023b. *Looking Back into the future: CRISPR and social values: BioGovernance Common's open discussion with Chinese academics*. Canterbury: Centre for Global Science and Epistemic Justice, University of Kent. https://blogs.kent.ac.uk/global-science-and-epistemic-justice/files/2023/02/Report-Looking-Back-into-the-Future-1.pdf

Zhang, Z. 2021b. *China ranks no. 1 globally in seven broad research areas*. China Daily. Available at: https://www.chinadaily.com.cn/a/202112/08/WS61b046fda310cdd39bc7a1d7.html

Webpage

GSEJ blogs on Biogovernance Commons 2022 Concluding Forums:
1) https://blogs.kent.ac.uk/global-science-and-epistemic-justice/2022/08/22/biogovernance-commons-concluding-forum-1/
2) https://blogs.kent.ac.uk/global-science-and-epistemic-justice/2022/08/22/biogovernance-common-concluding-forum-2/

Joy Y. Zhang is Professor of Sociology with a first degree in medicine. She is also the Founding Director of the Centre for Global Science and Epistemic Justice (GSEJ) at the University of Kent. An internationally recognised expert in the transnational governance of scientific risk, she specialises in emerging life sciences that impact both human and environmental futures. Her work advances sociological thinking on risk, cosmopolitanism, decolonisation, and subaltern politics, with a comparative focus on four major scientific powers: China, India, the UK, and the US.

Zhang is especially known for her expertise in science diplomacy: how scientific collaboration, governance, and public engagement can serve as sites for international negotiation, trust-building, and ethical coordination in times of geopolitical tension. In this regard, she has been at the forefront of efforts to connect responsible innovation, global governance, and diplomatic practice in the life sciences. She has initiated several landmark programmes, including the BioGovernance Commons (2021), the founding of GSEJ (2022)—whose research has informed G20 and G7 policy discussions—and, most recently, O.D.E.SS.I., a global 'odyssey' for public engagement with the life sciences. She is the author of three academic books and publishes widely in high-impact journals across both the natural and social sciences. She is frequently interviewed by global media and advises leading research and policy institutions in Europe and Asia.

Kathleen M. Vogel is Professor at the School for the Future of Innovation in Society and Senior Global Futures Scientist, Global Futures Laboratory at Arizona State University. Over the past ten years, she has conducted and published several research studies on biosecurity issues, including her 2013 book Phantom Menace or Looming Danger?:A New Framework for Assessing Bioweapons Threats (Baltimore: The Johns Hopkins University Press, 2013). In her research, she is generally interested in how knowledge is produced on controversial policy issues in the life sciences. She received her Ph.D. from Princeton University.

Sonia Ben Ouagrham-Gormley is an Associate Professor and the Deputy Director of the Biodefense Graduate Program at the Schar School of Policy and Government at George Mason University. She received her PhD in development economics from the Ecoles des Hautes Etudes en Sciences Sociales (EHESS) in Paris. Ben Ouagrham-Gormley has conducted research and written on such topics as nuclear, chemical, and biological weapons proliferation, organization and management of weapons programs, WMD trafficking in states of the former Soviet Union, biosecurity and bioterrorism, bio-dissuasion, export controls, defense industry conversion, transfer mechanisms of WMD expertise, and redirection of WMD experts.

Ruipeng Lei is Professor of Bioethics & Director, Center for Ethics and Governance of Science and Technology, and Vice Dean of Advanced Institute of the Humanities and Social Sciences, University of Electronic Sciences and Technology of China (UESTC) since 2023. She established Research Center for Bioethics at Central China University of Science and Technology in Wuhan in 2002 as Co-founder and Executive Director till 2022. Her research interest focuses on ethical and governance issues relevant to emerging technologies (Xenotransplantation, Human genome editing, Neurotechnologies, AI and Robotics, etc.). She has published over 100 research, review and commentary articles and 10 books written and edited in English and Chinese. She published a comment article "Reboot ethics governance in China" as first author in *Nature* in 2019 which has a positive impact on a series of regulations and legislation in China. Currently she has completed one National Key R&D project (2019-2024) and is conducting two National Key Projects regarding ethical and regulatory issues raised by synthetic biology, biobank and biomedical and health data as PI. She has been actively participating in national and international organizations including The Hastings Center, WHO Ethics and COVID-19 Working Group, UNESCO Working Group of the Draft Recommendation on the Ethics of Neurotechnologies, Asian Bioethics Association, Chinese Society for Bioethics, Chinese Society for Ethics of Science and Technology, Chinese Society for Synthetic Biology etc.

Chapter 2
Deconstructing Nationalist Science: Humanizing COVID-19 Research in the Era of Whole-Nation System

Lu Gao

Abstract This paper examines the complex landscape of scientific research in China during the COVID-19 pandemic, with a focus on how science is shaped by national policy frameworks and global dynamics. The study scrutinizes the new whole-nation system's role in shaping research directions and outcomes, while also highlighting the pressures faced by scientists from both domestic and international communities. Through interviews and case studies, the paper reveals the multifaceted nature of scientific work in China, including the impact of governmental policies, public perception, and international collaboration. It challenges the prevalent nationalist portrayal of scientific progress, arguing for a more nuanced understanding that appreciates the global interconnectedness and individual contributions within the scientific community. By foregrounding these dimensions, the paper proposes a "humanizing" perspective on science, emphasizing the role of uncertainty, communication, and situated judgment in knowledge production. In doing so, it offers a more nuanced account of how science operates in a context shaped by both global interdependence and geopolitical tension.

Keywords Nationalist science · Humanizing · COVID-19 · The new whole-nation system

Introduction

In 2021, as the COVID-19 pandemic unfolded globally, a group of scholars concerned with biogovernance found ourselves geographically dispersed yet intellectually connected through the BioGovernance Commons, an initiative proposed by Joy Zhang. Through sustained discussions across disciplines and regions, a persistent

L. Gao (✉)
School of Marxism, Tsinghua University, Beijing, China
e-mail: gao_lu@mail.tsinghua.edu.cn

J. Y. Zhang et al. (eds.), *BioGovernance Commons and China's Role in Global Risk Governance*, Emerging Globalities and Civilizational Perspectives,
https://doi.org/10.1007/978-3-032-32198-5_2

question emerged: how should we understand the role of science in a rapidly fragmenting global landscape, where knowledge production is increasingly entangled with national agendas and geopolitical tensions (Prasad 2023; Zhang And Burton 2022)?

During this period, a striking disjuncture became apparent. On one hand, the Chinese scientific community established its growing influence by sharing its findings. As Scopus shows, till Sept 2023, Chinese scientists have published 118,108 articles related to COVID-19 out of 955,504 in total, second next to the USA.[1] In response to the global outbreak of COVID-19, China moved quickly to produce and export PPE, ventilators, and medical supplies around the world. Despite seemingly positive initiatives, China's national image has worsened, even becoming less trustworthy in science (An et al. 2023). On the other hand, these contributions were often marginalized or selectively represented in global narratives, while domestically, the voices and lived experiences of scientists remained largely invisible. Existing accounts tended to frame Chinese pandemic science through the lenses of state mobilization or geopolitical competition, leaving little room to understand scientific practice as a situated, human process.

This paper responds to this gap by proposing a shift in analytical perspective. Rather than treating "nationalist science" as a stable or normative category, it approaches it as an analytical construct that requires unpacking. In particular, it asks: how is scientific work actually carried out within a state-organized system such as China's "whole-nation system" (juguo tizhi) and what forms of agency emerge within it? This question is especially salient in light of recent institutional transformations and policy orientations that emphasize coordinated national innovation while reshaping the conditions of scientific communication and visibility (Gan et al. 2020; Zhang And Barr 2020).

To address this question, I introduce the notion of humanizing science, which builds on recent discussions in the Journal of Responsible Innovation special collection Humanizing RRI in a Cosmopolitan World.[2] This line of work calls for a shift from macro-level institutional frameworks toward a more individual- and practice-oriented understanding of responsibility, emphasizing how scientific and technological governance is enacted through situated actors rather than abstract "stakeholders." By "humanizing," I do not mean simply adding individual stories to existing accounts, but foregrounding the practices, judgments, uncertainties, and negotiations through which scientists engage with both epistemic and institutional constraints. This perspective shifts attention from abstract narratives of state power or national competition to the micro-processes through which knowledge is produced, validated, and mobilized. In this sense, humanizing science reopens the

[1] https://www.scopus.com/results/results.uri?sort=plf-f&src=s&sid=9498b421dbb45eb18c2f0a8dfa65a955&sot=a&sdt=a&sl=8&s=COVID-19&origin=searchadvanced&editSaveSearch=&txGid=62d27051650f39bb10e6b92781300573&sessionSearchId=9498b421dbb45eb18c2f0a8dfa65a955&limit=10

[2] https://www.tandfonline.com/journals/tjri20/collections/Humanizing-RRI-in-a-Cosmopolitan-World

"black box" of scientific production by examining how structural arrangements are enacted and negotiated in practice.

Empirically, this paper draws on a series of in-depth interviews conducted between 2021 and 2022 with life scientists involved in COVID-19 research in China. These accounts reveal a more complex picture than that suggested by dominant narratives. While research was indeed embedded within a highly coordinated national system, it also involved significant degrees of uncertainty, improvisation, and transnational engagement. Scientists navigated tensions between urgency and rigor, visibility and caution, as well as national expectations and global scientific norms. These findings foreground the lived experiences of scientific work and highlight how agency is continuously negotiated within, rather than determined by, institutional arrangements.

These findings, however, are not unique to the Chinese context. Recognizing this complexity is essential not only for a more accurate account of China's pandemic science but also for rethinking how scientific knowledge is produced and governed in an increasingly fragmented world. We are witnessing a broader reconfiguration of science under intensifying geopolitical pressures, where concerns over national security, technological competition, and strategic autonomy are reshaping scientific systems across multiple countries (Prasad 2023; Zhang 2022a, 2022b). In this context, what is often described as "nationalist science" should not be understood as a phenomenon confined to a single national setting, but as part of a wider transformation affecting both established and emerging scientific powers, where issues of research security and techno-nationalism have become increasingly prominent (Rungius And Flink 2020). China, in this regard, represents one particularly visible and analytically significant case through which these broader dynamics can be examined.

The New Whole-Nation System and the State-Planned COVID-19 Research

In March 2023, China's Ministry of Science and Technology underwent a significant reorganization.[3] It reduced numerous project review and administrative tasks, transferring functions related to various industries to agencies such as the Ministry of Agriculture and the Health Commission. Relying on the Ministry of Science and Technology, the Central Scientific and Technological Commission (CSTS) was established. Its primary role is to provide the country with science and technology strategies and policies, to promote the *new whole-nation* system, and to optimize the management of the entire science and technology innovation chain.

[3] Proposal for the Reorganization of the State Council9https://www.gov.cn/xinwen/2023-03/11/content_5745977.htm?eqid=f2f686850005ffdd0000000464757ec6`

During the period dominated by the planned economy, the traditional whole-nation system-based practices were characterized by high centralization. In science, China reached a number of significant milestones through its whole-nation system, including the development of nuclear bombs, intercontinental missiles, and artificial satellites, as well as the synthesis of bovine insulin. Later, the country's reform and opening-up boosted its socialist market economy, enabling it to develop swiftly. In science and technology, focusing on crucial programs by pooling resources nation-wide shifted to institutionally ensuring the implementation of key national scientific research projects. The 863 and 973 Programs[4] presented a key model showing how scientific and technological resources from around the country were allocated to a program during the period.

By comparing it with the traditional whole-nation system, we can better understand the essence and "novelty" of the new whole-nation system. First, the mode of resource allocation seeks to return to the traditional "whole-nation" form. At the same time, "new" implies a breakthrough from the traditional planned economy model, integrating the forces of the market with the government and private with official entities. Some researchers believe this new whole-nation system represents a leftward shift in China's science and technology system and a climax in the changes in China's science and technology strategy over the past 4 to 5 years (Braun Střelcová et al., 2022). In fact, this shift has continuity in China's national science and technology policy over the past twenty years. At the beginning of the twenty-first century, China's science and technology system showed two trends: first, a lax approach where the evaluation system for academic research focused more on publishing papers than solving strategic problems needed by the nation; second, the unsuccessful commercialization of scientific research, where the forces of research were distracted by innovation and market incentives, making it difficult to form a united effort. The new whole-nation system aims to address these two issues and asserts its dominance over academia. In this matter, the Chinese government often benchmarks against the United States, particularly the success of DARPA in certain highly organized scientific projects. China feels that it has lost control over scientists, who are preoccupied with publishing papers. At the same time, major national tasks like semiconductor chip and large aircraft development often fail due to market disruptions and insufficient innovation-driven efforts.

Post-18th National Congress of the Communist Party of China in 2015, Xi Jinping at the Fifth Plenary Session of the 18th CPC Central Committee, explicitly proposed the idea of "exploiting the advantages of the new whole-nation system under market economy conditions, and hope to apply it to the research and development of central, key and basic technologies to break the bottlenecks and constraints

[4]The 863 program (863计划) was a program funded and administered by the Chinese Ministry of Science and Technology which intended to stimulate the development of advanced technologies in a wide range of fields for the purpose of rendering China independent of financial obligations for foreign technologies from 1986. 973 Program (973计划) or also known as National Basic Research Program is a basic research program initiated by the MOST to achieve technology and strategic edge in various scientific fields and especially the development of the rare earth minerals industry.

in respect of resources, technology, talents and other elements." Along with the ever-changing international climate, China started to face embargoes from the West in the high-tech area. The Fourth Plenary Session of the 19th CPC Central Committee again put forward the idea of building a new whole-nation system for tackling key core technologies under socialist market economy conditions (Li et al. 2023).

The outbreak of COVID-19 in Wuhan at the end of 2019 provided a training ground for the new Whole-nation system. On January 22, 2020, the day before the Wuhan lockdown, the first package of eight emergency research programs was launched by the Ministry of Science and Technology (MOST) as "scientific measures to address the pneumonia epidemic caused by novel coronavirus infections"[5]; five major technological routes for researching and developing COVID-19 vaccines were planned, including the inactivated vaccine, the genetically engineered recombinant subunit vaccine (recombinant protein vaccine), the adenovirus vectored vaccine, the nucleic acid-based vaccine (including DNA- and RNA-based), and the live attenuated influenza virus-vectored vaccine.[6] Each route required the synergy of several stakeholders, ranging from enterprises to universities, scientific research institutes, and high-level biosecurity laboratories.

The inactivated vaccine was a favored technical route based on China's long-standing research and production strength in this area. A 14-member expert group was established, led by Academician Zhong Nanshan, a pulmonologist recognized for his work in controlling SARS. This group also included virologists, epidemiologists, and traditional Chinese medicine experts. They served as the primary decision-making force, providing a scientific basis for China's COVID-19 research and policy decisions at various stages. According to interviews, the following issues were mainly taken into consideration when the R&D approaches were chosen: (1) the country's basic conditions and existing research basis. For example, given that China's research, development, and quantitative production chain of inactivated vaccines is the most mature, only integrating vaccine research with the vaccine industry can ensure the ultimate establishment of a comprehensive line of immunity defense. Therefore, taking into account the country's huge population and its existing research basis, the state decided that inactivated vaccines were the dominant route; (2) the self-sufficiency of vaccines. As China could basically accomplish an immunization plan on its own, the vaccine production and research should strive to acquire support from domestic technologies and production lines mainly and cover the dosages that the vaccine diplomacy might need (Meng et al. 2021); and (3) the state encouraged simultaneous development of different immunization plans and the competition among them, i.e., adopting the "horse racing mechanism." While boosting developments in emerging technological fields, the state ensured the early

---

[5] *The MOST and related parties jointly engage in emergency scientific research programs to address the COVID-19 pandemic*. https://www.most.gov.cn/kjbgz/202001/t20200123_151231.html. Accessed 2022-12-10

[6] Including the inactivated vaccine, the genetically engineered recombinant subunit vaccine (recombinant protein vaccine), the adenovirus vectored vaccine, the nucleic acid-based vaccine (including DNA- and RNA-based), and the live attenuated influenza virus-vectored vaccine.

implementation of those comparatively mature programs.[7] At the beginning of the pandemic, the Ministry of Science and Technology expected to allocate research resources nationwide and put different types of vaccines into its COVID control and prevention "toolbox." We can observe that the research and development related to COVID-19 in China is a state-led process, implementing the whole-nation system approach. Does this mean that there is limited scope for individual contribution and a lack of international elements of innovation within this whole-nation system?

Globalizing the Whole-Nation System

China's health diplomacy boasts a rich history, tracing its roots to the Mao Zedong era. Gaining prominence, particularly in the Global South, it has functioned as a pivotal foreign policy instrument, enabling China to cultivate an image as a responsible global power and further its geopolitical objectives. Over the past decade, Beijing has reevaluated and realigned its foreign and security policies, considering its growing political and economic clout, subsequently reshaping a variety of strategic interests.

In 2015, in a move that resonated with the Belt and Road Initiative (BRI), the National Health and Family Planning Commission unveiled the Three-Year Implementation Plan (2015–2017) for Advancing Health Exchange and Cooperation within the Belt and Road framework, heralding the inception of the Health Silk Road (HSR).[8]This period also witnessed the launch of Beijing's most comprehensive health aid initiative during the Ebola outbreak in Western Africa, signifying a pivotal shift in its health diplomacy and emphasis on scientific research.

Prior to the Ebola crisis, China's health diplomacy was predominantly characterized by dispatching medical professionals and providing basic medicines, with minimal impact on its direction in scientific research. The nation's scientific capabilities were then insufficient for tackling emerging infectious diseases. The Ebola outbreak, however, marked a turning point, thrusting scientific research to the forefront of China's health diplomacy initiatives. The subsequent development of COVID-19 vaccines underscored this significant paradigm change. Before 2014, China played a marginal role in Ebola research, as evidenced by the scant support from the National Natural Science Foundation of China for research literature (Yi et al. 2016). Post-2014, however, there was a notable shift towards aligning Chinese scientific endeavors with the production of global public goods. This era saw Chinese scientists diving deep into Ebola research in Africa and providing testing

---

[7] Interview with official(s) of the MOST, 2022-11-05.

[8] General Office of the National Health and Family Planning Commission of the People's Republic of China, "Notice on Issuing the Three-Year Implementation Plan (2015–2017) for Promoting Health Cooperation under the Belt and Road Initiative," October 14, 2015, https://www.nhc.gov.cn/zwgk/zxgzjh/201510/ce634f7fed834992849e9611099bd7cc.shtml (accessed December 10, 2022)

services, thereby embedding Chinese scientific pursuits within a more expansive international context (Larson 2014).

Three years on from the Ebola outbreak, six Ebola-related products in China were approved by the National Medical Products Administration (Li et al. 2020a). A team led by Dr. Chen Wei of the Academy of Military Sciences developed the world's first Ebola vaccine using a novel adenovirus genotype as the carrier (Cabestan 2021). This research laid the groundwork for one of China's five vaccine approaches to COVID-19 in 2020, demonstrating China's growing capacity to respond to global health crises and contribute to vaccine research and development for emerging infectious diseases.[9]China's ever-expanding "ambition" in politics has placed "Chinese science" into a broader international landscape and provided Chinese science with opportunities to produce global public goods.

> Compared to ten years ago, our perspective has become more comprehensive and diverse, contemplating how to transform research results into products. The scope of product application has broadened. Previously, we might not have considered it our responsibility to provide vaccines and products to Middle Eastern countries.[10]

Along with the implementation of China's BRI, trade and personal exchanges between China and the Middle East became increasingly frequent. MERS-CoV[11], another highly impactful coronavirus, which broke out in 2012 in the Middle East, outbreak in Korea in 2014 put a lot of strain on the global health system. Chinese researchers began to increase their efforts in the research and development of MERS vaccines (Lu et al. 2013). In August 2018, the research group in IMCAS (Institute for Microbiology, Chinese Academy of Sciences) took the receptor-binding domain ("RBD") of the MERS virus as the vaccine target. It transferred a complete set of technologies to Chongqing Zhifei Biological Products Co., Ltd. ("Zhifei Biological"), which started, in turn, the commercialization of the MERS vaccine.[12]

By the outbreak of COVID-19 in Wuhan in 2020, the cell line prepared for MERS had been fully completed.

> After the outbreak of COVID-19, we put our initial plan on hold. Knowing that the cause of the disease was a coronavirus, we immediately built a universal vaccine structure for β coronaviruses and published our results in Cell (Dai et al. 2020).[13]

The scientists continued their cooperation with Zhifei Biological and received approval for clinical trials in June 2020. This was one of the earliest recombinant protein vaccines that started its clinical trials worldwide. Later, the Phase I and II

---

[9] Interview of Scientist D, December 2021.

[10] Interview with Scientist Y

[11] In 2012, a respiratory disease caused by a corona virus was first discovered in Saudi Arabia. The disease featured a death rate as high as 37% and was mainly found in the Middle East, including Saudi Arabia and United Arab Emirates. The first patient with the disease was discovered in China in 2015. He was South Korean.

[12] Vaccine technology transferred from the IMCAS to Zhifei Biological. https://www.sohu.com/a/246051098_682259 Access 2022-11-10

[13] Interview with Scientist D, December 2021.

clinical trials of the vaccine were completed domestically. Given that limited cases existed in China for the Phase III clinical trial of the recombinant protein vaccine, facilitated by the Alliance of International Science Organizations ("ANSO"),[14] the Phase III clinical trial was launched in December 2020 abroad, such as in Uzbekistan.[15] After the clinical trials were completed, this vaccine was quickly put to use in Uzbekistan. In return, Uzbekistan could produce the vaccines and export them to neighboring countries.

Under the narrative of nationalism, as exemplified by the whole-nation system, global elements tend to recede or are less frequently mentioned in terms of their impact on work, thereby emphasizing national strength and capabilities. Another interesting case occurred at the IMCAS. A team from IMCAS worked out the JS016 antibody based on their research on the intrusion mechanism relating to the RBD of the coronavirus (Shi et al. 2020). The approval for the clinical trials was granted on June 5, 2020, making the antibody the first effective therapeutic antibody in the world to start its clinical trials. Researchers and Shanghai Junshi Biosciences Co., Ltd. ("Junshi Biosciences") have been jointly promoting the domestic clinical trials of the drug. In the end, Junshi Biosciences has licensed out the antibody for the overseas market to Eli Lilly and Company ("Lilly"). Patients around the world received one million doses of this therapeutic antibody in 2021, contributing a profit of USD 2.1 billion to Lilly.[16] Nonetheless, when talking about this achievement, a scientist expressed a sense of humor unique to the current era:

> Our institute is the only research entity in the world that has ever worked out both a vaccine and an antibody. We have licensed out the antibody to Lilly in the end because we appreciate the company's strengths in respect of research, development, and channels in the world. However, given the current relationship between the U.S. and China, we cannot publicize

[14] Within the BRI framework, the ANSO is the first comprehensive and substantive international science and technology organization jointly initiated and established by science research institutes, universities and international organizations in the countries covered by the BRI. The ANSO is committed to promote international cooperation in science, technology, innovation and capacity building in the countries and regions covered by the BRI, serves as an international cooperation platform for science to support the BRI construction and the sustainable development of global society and economy, and strives for contributing to build a community of shared future for mankind and helping reach UN sustainable development objectives. The founding members of the ANSO consist of the Chinese Academy of Sciences ("CAS"), the Russian Academy of Sciences ("RAS"), the Pakistan Academy of Sciences ("PAS"), Hungarian Academy of Sciences ("MTA"), the National Science and Technology Development Agency of Thailand ("NSTDA"), the National Academy of Sciences of the Republic of Kazakhstan ("NAS RK"), Tribhuvan University of Nepal ("TU"), Jomo Kenyatta University of the Agriculture and Technology ("JKUAT") and the United Nations Educational, Scientific and Cultural Organization ("UNESCO"). As of May 2020, the ANSO had had 52 members.

[15] Uzbekistan certifies Chinese vaccine to fight against COVID-19 - Xinhua | English.news.cn (xinhuanet.com) http://www.xinhuanet.com/english/2021-03/02/c_139778012.htm

[16] Eli Lilly and Company, *2021 Annual Report*, 2022, https://assets.ctfassets.net/srys4ukjcerm/W0elkCmxLN0Fmoisc1IVl/372392020400d3e57ed704b82257e3a3/Lilly_2021_Annual_Report.pdf (accessed December 10, 2022).

this accomplishment extensively. Interestingly, if the relationship between the two countries improves in the future, this may become a good story of international cooperation.[17]

One could argue that we're only a flight away from a novel virus, highlighting the rapid global spread and potential containment of diseases. However, the pandemic has also exposed the profound fragmentation and disparities within societies across different countries. These health disparities are stark reminders of our interconnected yet divided world. In contrast to the inseparability implied by global connectedness, every country grapples with significant health inequities within its borders.

Despite these challenges, the necessity for research and innovation transcends national policies and geopolitical tensions. No matter how isolated or adversarial countries may become, scientific endeavors—encompassing article publication, recognition of achievements, completion of experiments, product distribution, and vaccine diplomacy—must always be pursued with a global perspective. This approach is essential for addressing universal health challenges and advancing medical science.

Therefore, when we observe Chinese scientists through a magnifying glass, we find that their image is multifaceted, diverse, and international. For scientists, China's expanded political vision in the international arena has significantly inspired their imagination in scientific research. During interviews, some scientists mentioned that they used to follow the themes of scientific research more, but now they pay more attention to understanding social needs, which are not only the needs of China but also the international needs for emerging infectious diseases.

"The development of the Belt and Road Initiative has made us realize that the medical and health needs of many regions along the route have become our research topics, which is an expansion of our thinking."[18] Regardless of how China's internationalization strategy's political intentions are evaluated, from the perspective of global scientific health development, Chinese science serving global public goods is promising and may even become an important force to balance political contention.

Humanizing COVID-19 Research in a Mutually Skeptical Global Arena

Since the spread of the COVID-19 pandemic, the virus has not only been labeled as the "Wuhan virus," but Western media have also criticized China for hiding secrets on many issues, such as virus origin and transparency of information (Hernandez And Gorman 2021). Meanwhile, the Chinese government and social media have been propagating the theory that the virus originated from a U.S. military lab,

---

[17] Interview with Qian. 20 Dec 2022

[18] Interview with

emphasizing issues of biosecurity.[19] Under the influence of the political arena, both China and the West have fallen into a state of extreme distrust, and scientists are under immense pressure in such an environment.

In an interview published on *Scientific American* (June 2020), Shi Zhengli honestly admitted that she had doubted a lab leak and had checked the experimental data. The investigation showed that the genome sequence of the COVID-19 virus did not match any virus in her lab (Qiu 2020). However, her candidness in the interview became a subject of media skepticism. Doubts about the lab leaking did not diminish because of her self-examination, but instead grew stronger. In a report published in *The Independent* in June, 2021， Shi said, "How on earth can I offer up evidence for something where there is no evidence?"[20] The media's widespread promotion of the toxic China theory has led to distrust in Chinese science and a lack of recognition for Chinese scientists in the West, prompting a large number of Chinese scientists to return to China. As Joy Zhang said in her new book, the global politicization of COVID seems to be in danger of playing into China's nationalist narratives (Zhang And Burton 2022).

Inside China, scientists also navigate institutional constraints on publication and communication, alongside growing challenges of public trust. January 29, 2020, *The New England Journal of Medicine* published an article titled "Early Transmission Dynamics in Wuhan, China, of Novel Coronavirus–Infected Pneumonia," authored by researchers from the Chinese Center for Disease Control (CCDC), the University of Hong Kong, and other institutions (Li et al. 2020a, 2020b). Based on epidemiological analysis of 425 laboratory-confirmed cases reported as of January 22, the study suggested that human-to-human transmission among close contacts had likely occurred since mid-December and gradually expanded thereafter. As one of the earliest retrospective analyses of the outbreak, the study attracted widespread public attention.

The reception of this finding, however, revealed a gap between scientific reasoning and public expectations. Retrospective epidemiological analysis, which reconstructs transmission dynamics based on accumulated data, was not always easily understood outside scientific communities. Earlier public communications from local health authorities in early January had emphasized the absence of clear evidence for sustained human-to-human transmission, reflecting the uncertainties at that stage. When later findings appeared to suggest an earlier onset of transmission, this temporal disjuncture contributed to confusion and debate in the public sphere.

Around the same time, on January 31, 2020, the Ministry of Science and Technology (MOST) issued a notice on strengthening the management of COVID-19-related research projects. The document emphasized that scientific

[19] Embassy of the People's Republic of China in Spain, "Full Text Title (translated if needed)," March 23, 2020, https://es.china-embassy.gov.cn/chn/mtjj/202003/t20200323_3230865.htm (accessed December 10, 2022)

[20] Chantal da Silva, "China's 'Bat Woman' virologist speaks out over Covid lab leak theory," The Independent, June 15, 2021, https://www.independent.co.uk/asia/china/covid-wuhan-lab-leak-china-b1865627.html (accessed December 10, 2022).

research should prioritize epidemic prevention and control, and that research outcomes should primarily serve public health needs rather than immediate publication. The temporal proximity between emerging scientific findings and evolving policy guidance further shaped the context in which scientific work was interpreted.[21] This statement triggered an explosive reaction online.

George Gao, the director of the CDC at the center of the storm, faced severe online violence, with his personal information being exposed and even threats of physical attacks. He was also unjustifiably insulted for being trained as a veterinarian during his undergraduate studies and not being qualified as the director of CCDC. In fact, GAO is a typical top contemporary Chinese scientist who graduated from Oxford University in the 1990s, later taught at Oxford, and returned to China in 2004 to serve as the director of the Institute of Microbiology, CAS. This generation of scientists had the chance to receive the best education in the West and joined the ranks of world-class scientists. As a result, he is not only an academician of the Chinese Academy of Sciences but also a foreign fellow of the National Academy of Sciences in the USA and the Royal Society in the UK. At the same time, influenced by the traditional education from their parents, they possess a strong ambition to "serve the motherland." In 2014, he played a significant role in fighting the Ebola virus in West Africa.

Over time, as more information became available, the nature of the study and its conclusions were gradually better understood. The article in question was a retrospective epidemiological analysis, drawing inferences from cases that had already been identified. The conclusion regarding possible human-to-human transmission in mid-December was therefore not an immediate observation, but a reconstruction based on data available only in late January. This temporal gap between data generation and scientific inference contributed to misunderstandings about the timing and meaning of the findings. More broadly, the episode highlighted the differences between scientific reasoning, policy communication, and public interpretation. Scientific language, which often relies on probabilistic inference and retrospective reconstruction, does not always align with the expectations for clarity and immediacy in policy statements or public communication. As these different forms of expression intersected under conditions of urgency, discrepancies between them became more visible and, at times, difficult to reconcile.

Within this context, scientists found themselves navigating not only technical challenges but also heightened public visibility. Some researchers, including senior figures involved in the study, became focal points of attention and debate. While such scrutiny reflected the significance of their roles, it also had implications for ongoing research work. As one member of the research team noted in an interview, "If it weren't for this controversy, our team would likely have produced more—and possibly better—publications." This remark points to the ways in which broader social dynamics can shape the conditions of scientific practice, even as research continues to evolve.

---

[21] https://www.most.gov.cn/kjbgz/202001/t20200129151264.html Access 2022-11-10

Following the damage to Gao's credibility as the CDC director, Yan, a gentle and intellectual woman and Gao's key deputy, became the only scientist appointed by the State Council to announce national initiatives on COVID-19 medications and related scientific research.[22]

On February 17, the scientists in Gao's group obtained, for the first time, the structure of the angiotensin-converting enzyme 2 ("ACE 2"), an essential protein in the coronavirus, by adopting a structural biology approach. However, under the pressure of public opinion and the policy in China at the time, he chose to publish his data on the website of the National Microbiology Data Center instead of publishing a paper first (Wang et al. 2020). He said,

> More than 7000 downloads were made from 110 different IP addresses around the world within one month.[23] Choosing to share data first and then publish their papers became a stopgap for some Chinese scientists under the circumstances at the time.[24]

In May 2020, at an online seminar co-hosted by Tsinghua University, Beijing Frontiers of Biological Structure Center, and Science magazine, Gao boldly proposed "Seeking Answers from Science to the Challenges of COVID-19."[25]The spirit of scientific truth-seeking also became a key theme in many of his statements during the pandemic.

After the controversy, Gao's much-debated paper became evidence of China's achievement of pandemic resistance, displayed in the Museum of the History of the Communist Party of China, which opened in 2021.[26] In fact, the scientists I interviewed have always been aware that as members of the international scientific community, their most important responsibility is to publish their research findings in a timely manner. These contributions are continuously advancing the progress of science both in China and globally.

> Quality publications are essential in advancing overseas clinical trials and have played a decisive role in determining whether public health authorities of another country allow us to engage in clinical trials on our vaccines. To promote clinical trials, we negotiated with more than ten countries and finally chose Uzbekistan, Pakistan, Indonesia, and Ecuador for our Phase III clinical trials.[27]

[22] At 3:00 p.m. on February 15, the Joint Prevention and Control Mechanism of the State Council held a press conference at the Medium Conference Hall on the third floor of the East Building of the Guo'erzhao Hotel， https://www.gov.cn/xinwen/2020-02/15/content_5479345.htm

[23] Interview with QI.

[24] National Microbiology Data Center, "Global Coronavirus Genomics Data Sharing and Analysis System," https://nmdc.cn/zh_coronavirusCoronavirus Linked Data The RDF Knowledge-based Database for Coronavirus Researches (nmdc.cn). https://nmdc.cn/zh_coronavirus

[25] The Great Challenge of COVID-19: Seeking Answers from Science," Sina Live, https://zhibo.sina.com.cn/news/ty40008901

[26] It is displayed in the Museum of the History of Communist Party of China.

[27] Interview, Dai

Without the high-quality papers, it would not have been possible for China to start any clinical trials abroad or for Chinese medicine to go global. Then, the goal of providing public good as a responsible nation would be mission impossible.

We attempt to humanize COVID-19 research, thus revealing the multiple pressures faced by scientists in China. This includes skepticism from Western countries about Chinese science, distrust from the government, and questioning by the Chinese public. In our case, rebuilding trust heavily relies on scientists acting as members of the global scientific community, engaging in candid exchanges with their international peers. We might speculate that international publication becomes a key channel for Chinese scientists to preserve data and express their scientific viewpoints, and these opinions might, in turn, influence national policies through international channels. State-led science policies, similar to the whole-nation system, can efficiently concentrate resources and efforts to tackle difficult problems. However, respecting scientific facts in the decision-making process and abandoning strategies that prioritize national interests and social stability seems equally crucial. Meanwhile, with the current mutual skepticism between China and the West, science, and the scientific community, as a well-organized world system or Cosmopolitan, should play a more active role.

Conclusion

This narrative of China's scientific research development during the COVID-19 period might be portrayed as follows: China achieved numerous victories in COVID-19 research and development, attributed to the effective leadership of the government and the superiority of the whole-nation system. Under this system, scientists efficiently fulfilled their research and development tasks. Such a story may emphasize the central role of the state and the efficiency of centralized management, portraying scientific progress as part of national power and strategy. However, this narrative could lead to distrust from the West and the public, as it might be seen as overly emphasizing nationalism and overlooking international cooperation, openness, and individual autonomy in scientific research. Therefore, deconstructing the "nationalist" narrative might lead to more complex and comprehensive conclusions. The aim of this article is to present a more balanced and diverse perspective, exploring the real complexity of China's scientific research development during the COVID-19 period.

To confess, I was drawn to the topic of COVID-19 research by the aspiration for China's scientific rise. When I saw on social media the significant scientific discoveries and translational research made by scientists around me, I was very eager for more people to know about and even praise these works and China's scientific achievements. On the surface, these achievements seemed to have been accomplished within the whole-nation system during the pandemic. However, this research shows that scientific advancements result from the solid foundation of research accumulated in the last few decades, unfolding through cumulative and iterative processes, and also heavily gained from international collaboration, academic freedom, and the innovative spirit of individual scientists.

The research elucidates a salient tension inherent within the realm of science in China. This tension arises from the imperative of accommodating practical needs for transparent collaboration in the international scientific arena, the emergent mandate to articulate China's narrative effectively, and the intricate interplay of self-reflexivity and institutional governance that is pivotal for professional advancement within China's academic and research institutions. Deconstruction reveals that science should not be seen merely as a tool for national interests but as a global, cross-cultural effort aimed at addressing common challenges faced by humanity. Effective scientific practice relies not only on hard power but also on the adept use of soft power. Thus, a pivotal objective is to amplify Chinese narratives to a broader audience, ultimately bolstering the soft power of science.

When we realize the cultural value of science, as well as the fact that science also implies a globalized force that can eliminate particularities, as far as China is concerned, the understanding and shaping of the development of science becomes more important. By closely observing scientists covered by this article, we see that many cases in which scientists reconsider the value of their research: angriness about the local industry's failure of doing its best to compete with foreign industries, persistence in making research achievements, criticism to the innovation culture, expectation for reforms, dissatisfaction with the political environment, perseverance with diligent practices under various pressures, and ideals of sticking to the spirit of ancient Chinese scholars and building a perfect world. Perhaps the term "consciousness of culture," introduced by the Chinese anthropologist and sociologist Fei Xiaotong (2003), can be borrowed here to describe these Chinese scientists, who are entering a state characterized by a consciousness of scientific culture. These scientists are gradually acquiring a kind of "self-knowledge" and "consciousness" and keep reconsidering the complicated political, cultural, and international environments in which they do their jobs. In such interactions with the environment, Chinese scientists more soberly realize where the value of scientific research itself lies than in the past, and gain an independent position that fits in with the future. We call such a phenomenon the consciousness of the scientific culture. Over decades, Chinese scholars have been used to looking towards the West, finding beauty in differences between others and themselves, and integrating such an understanding with their self-reflections. The course of China's modernization over the past nearly one hundred years could be regarded, to some extent, as a course in which China has been globalized. Nevertheless, what is more important for the future is to present this consciousness courageously to the world, promote a kind of "reciprocal observation and interaction," and shape an international relationship and world structure, in which countries appreciate their own strengths and others' alike and focus on harmony while respecting differences.

References

An, N., Z. Zheng, C. Chen, et al. 2023. Mapping a country image from global news reports about COVID-19 pandemic. *Applied Spatial Analysis* 16:751–770.

Braun Střelcová, A., S. Christmann-Budian, and A. L. Ahlers. 2022. The end of 'learning from the west'? Trends in China's contemporary science policy. *Observations* 6, Max Planck Institute for the History of Science. https://doi.org/10.17617/2.3407491.

Cabestan, J.-P. 2021. China's response to the 2014–2016 Ebola crisis: Enhancing Africa's soft security under Sino-US competition. *China Information* 35 (1): 3–24.

Dai, L., T., Zheng, K., Xu, Y., Han, L., Xu, E., Huang, et al. 2020. "A Universal Design of Betacoronavirus Vaccines against COVID-19, MERS, and SARS." *Cell* 182 (3): 722–733. https://doi.org/10.1016/j.cell.2020.06.035

Fei, Xiaotong. 2003. Some Confessions on 'Cultural Consciousness. *Academic Research* 7:5–9.

Gan, N., C. Hu, and I. Watson. 2020. Beijing tightens grip over coronavirus research, amid US-China Row on Virus Origin. *CNN*. https://edition.cnn.com/2020/04/12/asia/china--coronavirus-researchrestrictions-intl-hnk/index.html.

Hernandez, J. C., and J. Gorman. 2021. On W.H.O. Trip, China refused to hand over important data. *New York Times*.https://www.nytimes.com/2021/02/12/world/asia/china-world-health--organizationcoronavirus.html.

Larson, C. 2014. *China ramps up efforts to combat Ebola*. Science. https://www.science.org/content/article/china-ramps-efforts-combat-ebola.

Li, C., J. Y. Chen, and Y. M. Huang. 2020a. Challenges and opportunities for China entering global Research and Development for emerging infectious diseases: A case study from Ebola experience. *Infectious Diseases of Poverty* 9:27.

Li, Qun, Xuhua Guan, Peng Wu, Xiaoye Wang, Lei Zhou, Yeqing Tong, Ruiqi Ren, et al. 2020b. Early transmission dynamics in Wuhan, China, of Novel Coronavirus–Infected Pneumonia. *New England Journal of Medicine*. https://doi.org/10.1056/NEJMoa2001316.

Li, Rui, Liang Zheng, and Lan Xue. 2023. The taxonomy and boundaries for the new whole nation based on the technological evolution theory. *Studies in Science of Science*. https://doi.org/10.16192/j.cnki.1003-2053.20230830.003.

Lu, G., Y. Hu, Q. Wang, J. Qi, F. Gao, Y. Li, et al. 2013. Molecular basis of binding between novel human coronavirus MERS-CoV and its receptor CD26. *Nature* 500 (7461): 227–231.

Meng, Z., S. Shan, and R. Zhang. 2021. China's COVID-19 vaccination strategy and its impact on the global pandemic. *Risk Management and Healthcare Policy* 14:4649–4655.

Prasad, E. 2023. *The future of money: How the digital revolution is transforming currencies and finance*. Cambridge, MA: Harvard University Press.

Qiu, Jane. 2020. How China's 'Bat Woman' Hunted Down Viruses from SARS to the New Coronavirus. *Scientific American*, June 1. https://www.scientificamerican.com/article/how-chinas-bat-woman-hunted-down-viruses-from-sars-to-the-new-coronavirus1/.

Rungius, C., and T. Flink. 2020. Romancing science for global solutions: On narratives and interpretative schemas of science diplomacy. *Humanities and Social Sciences Communications* 7:102.

Shi, R., C. Shan, X. Duan, Z. Chen, P. Liu, J. Song, et al. 2020. A human neutralizing antibody targets the receptor binding site of SARS-CoV-2. *Nature* 584 (7819): 120–124.

Wang, Q., Y., Zhang, L., Wu, S., Niu, C., Song, Z., Zhang, et al. 2020. Structural and Functional Basis of SARS-CoV-2 Entry by Using Human ACE2. *Cell*, 181(4), 894–904.e9. https://doi.org/10.1016/j.cell.2020.03.045

Yi, F., P. Yang, and H. Sheng. 2016. Tracing the scientific outputs in the field of Ebola research based on publications in the web of science. *BMC Research Notes* 9:221.

Zhang, Joy Yueyue. 2022a. *The Cosmopolitanization of science: Stem cell governance in China*. Cham: Palgrave Macmillan.

Zhang, Joy Yueyue. 2022b. The hegemonic paradox of science diplomacy and its contemporary challenges: Lessons from theCOVID Pandemic. *Science Diplomacy Review*. https://www.ris.org.in/journals-n-newsletters/Science-Diplomacy-Review.

Zhang, Joy Yueyue, and Datta Burton. 2022. *The elephant and the dragon in contemporary life sciences: A call for Decolonising global governance*. Manchester: Manchester University Press.
Zhang, J., and M. Barr. 2020. Harmoniously denied: COVID-19 and the latent effects of censorship. *Surveillance & Society.* 19 (3): 389–402.

Lu Gao is an Associate Professor at Tsinghua University. Before joining Tsinghua, she was an Associate Professor at the Institute for the History of Natural Sciences, Chinese Academy of Sciences, where she also served as Director of the STS Center. She has been a Visiting Scholar at the University of Kent, Stanford University, and the University of Edinburgh. She serves on the editorial boards of East Asian Science, Technology and Society and Journal of Responsible Innovation, and holds leadership roles in several major Chinese academic associations in science policy, technology governance, and ethics.

Her research lies in Science, Technology and Society (STS), with a focus on the governance of emerging technologies, particularly biotechnology and AI-driven science. She examines how ethical and governance frameworks are historically formed through the interaction of scientific practice, institutional arrangements, and public values. Combining STS, history of science, and science policy approaches, her work analyzes gene editing, genomics, synthetic biology, and dual-use life sciences, while comparing Western governance experiences with Chinese pathways. She has published over 40 academic articles in leading journals such as Journal of Responsible Innovation, AI & Society, Cultures of Science, and East Asian Science, Technology and Society. Her recent work also explores AI for Science and the transformation of laboratory life, and contributes to conceptual developments such as "Humanizing RRI." She is the author of the forthcoming monograph From Participation to Co-Governance (2025) and has led multiple national and international research projects.

Chapter 3
Afraid of Whom? A European Outlook on China in Between Research Priorities and Biographical Notes on the Encounter with the Other

Daniela Caterina

Abstract This chapter explores how Europe perceives and engages with China against the backdrop of global crises and shifting geopolitical orders. It argues that dominant China discourses—often monolithic, polarised, and Anglophone—obscure the complexity of China's rise in S&T while reinforcing binary 'China versus the rest' framings. Drawing on Gramsci's dialectic of philosophy and politics, the analysis situates debates on China's technological ambitions, academic cooperation, and research governance within wider reflections on knowledge production and ethical responsibility. Using both a macro-political and biographical lens, the chapter connects European research priorities to lived experiences of academic migration and intellectual encounters with the Chinese Other. Ultimately, the chapter calls for a complexity turn in European approaches to China—one that restores reflexivity, interdisciplinarity, and humanity in cross-cultural engagement, even as global crises such as the Covid-19 pandemic, the Russian invasion of Ukraine, and the war in Gaza continue to test our capacity for ethical global dialogue.

Keywords Sino-European relations · China images · Academic migration · Gramsci · Knowledge geopolitics · Complexity

Introduction

The 20th Congress of the Communist Party of China (CPC) in October 2022 reaffirmed the centrality of science and technology (S&T) in China's aspiration to become a global power. In particular, the primacy that Xi Jinping places on S&T as the country's 'primary productive force', on talent as its 'primary resource', and on

D. Caterina (✉)
School of Philosophy, Huazhong University of Science and Technology, Wuhan, China
e-mail: caterina@hust.edu.cn

J. Y. Zhang et al. (eds.), *BioGovernance Commons and China's Role in Global Risk Governance*, Emerging Globalities and Civilizational Perspectives,
https://doi.org/10.1007/978-3-032-32198-5_3

innovation as its 'primary driver of growth' implies profound economic ramifications—as commentators anticipated with an eye to China's sustained 'epic investment in science' (Mallapaty 2022). Since then, empirical evidence has reinforced and expanded this narrative: for example, China's national innovation index rose markedly in 2023, driven by rapid growth in the innovation environment and output indices (National Bureau of Statistics of China 2023). At the same time, recent scholarship shows that China's S&T system is undergoing a governance shift: the reform of the institutional architecture (e.g., establishment of the Central Science and Technology Commission in 2023) is intended to tighten centralised control while re-balancing central-local dynamics (Xiao and Xu 2024). Additionally, researchers point to the emergence of a 'whole-of-nation' innovation logic in China's system, combining large-scale state direction with strategic priorities in key technologies and emphasising self-reliance (Groenewegen-Lau 2024).

However, and crucially for our argument, China's emphasis on S&T cannot simply be reduced to macroeconomic indicators or confined to natural scientists working in growth sectors. A social science lens is essential to grasp what is at stake. In the life sciences, for instance, China's biotech or biomedical ambitions inevitably raise difficult issues, from the benefits of growth to questions of political influence, securitisation, ethical oversight, and global governance. These challenges are further mediated—and often intensified—by political differences between China and other major scientific powers, particularly in relation to governance models and state-market relations, as well as norms of transparency, accountability, and individual rights. Such differences shape not only how scientific collaboration is pursued and regulated, but also how risks are perceived, how trust is negotiated, and how standards of research security and ethics are defined and contested across borders. Indeed, the transformation of China's S&T system is deeply entwined with geopolitical dynamics, global knowledge flows, and norms of research security (Sun et al. 2025), all of which are filtered through—and complicated by—these political divergences. And then: *what* is meant by 'China' in all this conversation? What are the imaginaries, the institutional practices, the regional variations, the social actors, the networks, and the global entanglements behind the shorthand? A self-reflexive social science lens is therefore indispensable if we are to unpack not only these complex entanglements, but also how political difference operates as a structuring and mediating force within them.

In the following, I will ground this argument in two steps—in between the definition of key research priorities in our approach to China and biographical notes on the very emergence of this research agenda. First, I will sketch out the much broader picture in which to embed reflections on China's ambitions and advancements in S&T, arguing for a critical assessment of the lights and shadows of dominant China debates in an increasingly heated geopolitical context. Second, drawing on my specific interplay of personal and professional experiences, I will touch upon how this macro-level of analysis is inherently related to our 'human story' (Zhang et al. 2022, 48) as scientists and, at a more fundamental level, as human beings engaging in real-life China encounters. Taking stock of these considerations, I will eventually conclude my perspective on the encounter with the Chinese Other by articulating a call to restore complexity in Sino-European interactions—in science and beyond.

Afraid of (which) China?

'China is, simultaneously, in different policy areas, a *cooperation partner* with whom the EU has closely aligned objectives, a *negotiating partner* with whom the EU needs to find a balance of interests, an *economic competitor* in the pursuit of technological leadership, and a *systemic rival* promoting alternative models of governance' (European Commission 2019, 1, emphasis added). Cooperation partner, negotiating partner, economic competitor, systemic rival: from a European perspective, relations between China and the European Union (EU) continue to be stuck in the ambivalences of the 2019 Strategic Outlook, which pleads for a 'flexible and pragmatic whole-of-EU approach' (ibid.) to address the *many faces* of the Chinese Other. It goes without saying that China's breathtaking emergence as a global power in S&T can hardly be separated from this bigger picture. Either in this specific strategic field or regarding China's global rise more in general, however, the tone of dominant China debates—in public discourse as well as in academia—is strikingly far from what the call for a '*multi-faceted* approach to China' (ibid., emphasis added) might suggest.

We thus seem to be stuck in a paradox. On the one hand, references to China abound like never before. 'China' has become more important than ever in our daily lives—we have never been so interested in what China decides, what it communicates, supports, rejects, produces, invests in, sells, and buys. To be sure, as put by Vukovich, 'it is not that we now know China better, but that the global economic footprint of China is too large not to enter our consciousness and to produce more "China references"' (Vukovich 2010, 164). Yet, *what* is this China *we* refer to? *Who* is China in so much ubiquitous, controversial talk? On the other hand, China references, namely, turn out to be outstandingly thin. They tend to reproduce and consolidate a perception, representation, and understanding of 'China', first, as an alleged unitary, monolithic actor; second, in outright opposition to an (equally unitary, monolithic) 'rest of the world'; and third, coupled with an understanding of the 'West' as exclusively US American or, at best, Anglophone. To put it in simple terms, the paradox is that the more China consolidates its stance as a global player, the more 'China' threatens to remain a black box to external eyes.

A look at the political science debate on China's rise is highly telling in this respect, as it is symptomatic of these three main limitations plaguing China debates more generally. The first two limitations—(i) monolithic and (ii) polarised/dichotomic China references—are paradoxically related to the wealth of (dissenting) voices in 'China watching' (Pan 2012). Empirical evidence about the features, motives, and trajectories of China's global rise is complex and even contradictory. Does China aim to reverse the dominant world order in a sustained attempt to supplant the USA? Or is China rather interested in preserving the status quo, as it is the first actor to profit from the present USA-led balance of power? Among scholars of international relations, disagreement on these issues is so endemic that they tend to disagree not only among various schools but even *within* their respective—realist, liberal, constructivist, or Marxist—camp (for an overview, cf. Jones and Hameiri 2021, 4–8). Such disagreement does not stop at analytical efforts at *understanding* China's rise but rather spills over to programmatic appraisals of how to *react* to such a rise. Should we refrain from finding similarities between the Chinese case and any other case in the light of the first's 'exceptional' nature? Should we adopt a 'maieutic' approach to support China's transition to a free-market liberal democracy? Should we rather adhere to 'whataboutism', thus diverting the focus on the fundamental flaws of our own political systems and dismissing any moralising attempt to criticise China as hypocritical? To be sure, as Franceschini and Loubere (2022, 1-8) remark, no clear-cut boundaries delimit adherence to such 'exceptionalist', 'maieutic', or 'whataboutist' frames, with shifts and overlaps unveiling endemic disagreement also over appropriate reactions to China's rise. Yet, such overall disagreement in 'China watching' is even more striking because of its paradoxically unitary effects. Despite uncountable internal divisions, in fact, dissenting voices in China analyses eventually work in the same direction, as they suffer from oversimplifications that tend to consolidate the first two dangerous trends pointed out above: first, a conception of 'China' as a monolithic political and socioeconomic force, and, second, a binary vision of 'China against *the rest*'—a concept that in a vicious circle, as Pavlićević & Talmacs (2022, 2) denounce, also suffers from oversimplification, thus ushering into a monolithic entity itself.

Adding to this, the third main limitation in academic and public debate, attempts at focusing on China from an explicit 'Western' perspective turn out to be largely subsumed to a US American or, at best, Anglophone standpoint. To be sure, Sino-US geopolitical tensions continue to play a crucial role, even more in the shadow of an alleged enduring 'New Cold War' (Brands & Gaddis 2021). However, it would be a methodological and substantial mistake to dismiss this Anglophone bias as either irrelevant or merely related to the present geopolitical conjuncture. A deep-going examination of the state-of-the-art concerning the study of Western perspectives on China at the crossroads of political science, language, and communication studies, in fact, reveals the same Anglophone bias in investigations focusing both on past and more recent China images emerging in the West (cf., e.g., Isaacs 1958; Iriye 1967; Mackerras 1989, 2013). Importantly, the overwhelming reliance on Anglophone sources is often coupled with two further trends in the literature: on the one hand, the almost exclusive attention devoted to perceptions and representations of China in (Anglophone) Western *media* to the detriment of a broader/different range of sources (cf., e.g., Yan 1998; Seib and Powers 2010; Zhang 2010); on the other hand, an overwhelming *constructivist* theoretical grounding preventing an in-depth investigation of how (Western) discursive representations and images of China dialectically relate to key extra-discursive, socioeconomic dynamics (cf., e.g., Pan 2012; Turner 2014). Awareness of these trends is key to gauge the enduring difficulties to stage a solid and productive cross-disciplinary dialogue between China studies and postcolonial studies (cf. Meinhof et al. 2017).

Far from representing technicalities for academic niche debates, I argue, the main limitations pointed out above in relation to the consolidation and capillarity of *thin* China references are of primary importance, as they bear significant consequences in our societies. First, abridged monolithic approaches to 'China' as an allegedly unitary actor fall short of illuminating the multi-layeredness of Chinese society—as any other society—in socioeconomic, political, and cultural terms. This way, they prevent an adequate understanding of and engagement with the complexity of its actors, structures, and discourses. Second, polarising dichotomic generalisations and simplifications playing 'China against the rest' force 'China'—and 'the rest'—to remain a black box or, worse, a collection of stereotypes. Third, from a European standpoint of observation, the overwhelming identification of a Western approach to China as US American/Anglophone forecloses a highly precious space to develop a specifically European strategy able to untap the potential of a more nuanced outlook in Sino-European relations.

The urgency of these considerations becomes even more tangible if we relate them to debates over specific policy priorities China's rise, such as the country's emphasis on S&T, as pointed out in the beginning. Sino-European interactions in this field are emblematic in this respect because they clearly show the (deleterious) short-sightedness of perpetrating *thin*—monolithic, polarised, culturally/linguistically biased—China debates. Indirectly, Zhang et al. (2022) provide extensive evidence for each of the three limitations to the debate discussed above. What is the point in addressing 'China' and 'Chinese science' or 'Chinese business' in monolithic terms? Who gains in overemphasising an alleged opposition between 'China' and 'Chinese scientists' or 'Chinese business' and 'the rest'? How much goes lost in unduly subsuming the complexity and richness of that 'rest' interacting with the Chinese Other—scientists, businessmen, citizens—in exclusively US American/Anglophone terms?

Taking stock of these considerations, I argue that to find a way out of abridged, polarised, and biased generalisations in Europe-China relations is a topical, urgent, and feasible task. To be sure, however, this task should not degenerate into naïve wishful thinking. Alerting public opinion towards a more nuanced and self-reflexive approach to China cannot—and does not aim to—compete with dominant scholarly, media, and policy discussions by completely subverting them. I namely agree with Franceschini and Loubere when they warn that fostering a more fine-grained, empirically grounded approach to China 'will hardly produce a narrative as appealing as those put forward by the proponents of 'debt traps', 'silent invasions', or benevolent 'win-win' rhetoric and 'South-South co-operation'; however, such an approach 'might get us closer to understanding what global China really means *for the people who experience it in their everyday life* and help us to better visualise the broader implications of Chinese globalization going forward' (Franceschini and Loubere 2022, 48, emphasis added).

Bringing the People Back in: A Personal China Encounter

With an eye to Europe-China relations, looking at the micro-level of scientists' interactions represents a promising entry point to zoom in on *what China really means for the people who experience it in their everyday life,* as freely quoted from Franceschini and Loubere above. To be sure, high-skilled European academic migrants moving to China represent a very limited group and could hardly be representative of overall Sino-European encounters. However, I argue, they play a crucial role in this respect. As put by Braun Střelcová and colleagues, '[i]n a way, researchers, both European researchers in China and their Chinese counterparts in Europe, are "science diplomats" who bridge their employing institutions and lay the *human foundations* for the relationship between Europe and China in higher education, research, innovation, *and beyond*' (Braun Střelcová et al. 2022, 24–25, emphasis added).

In light of China's policy priorities in the field of S&T, the topic of academic migration to China—in terms of a professionally motivated migration to China to settle there in a permanent way or, at least, more than 1 year (Bauder 2020)—is topical but still limitedly explored. On the one hand, China-bound academic migration has rarely been a main concern in studies of skilled migration, whose focus is predominantly on the political framework of such cross-border moves rather than on personal experiences (Zweig and Wang 2013). The findings of literature on internationalisation and mobility in higher education, in turn, can hardly be representative of the bigger picture of European academic migration to China, as they mostly focus on Sino-foreign joint venture universities (Cai and Hall 2016; Wang and Chen 2021). On the other hand, the literature in China studies has dealt with Europeans' academic migration to China in a mainly implicit way, as several sources offer an indirect outlook on the features (and difficulties) of China-based fieldwork or collaborations (Heimer and Thøgersen 2006).

Against this background, pioneering studies like Braun Braun Střelcová et al. (2022), which try to bridge these gaps both at the theoretical and empirical level, are most welcome. The innovative conceptual take of this research merges push-pull factors, Bourdieu's four-fold conceptualisation of economic, social, cultural, and symbolic capital, as well as Marx's concept of valorisation. In turn, at the empirical level, the triple focus on the motivations, job satisfaction, and career prospects of European China-bound academic migrants across the natural and social science divide represents a precious first step to open the black box of China-based European researchers' experience, while also fostering a more nuanced analysis of this increasingly relevant phenomenon within China's race for talents in S&T.

In my specific case, studies such as Braun Braun Střelcová et al. (2022) necessarily trigger a wave of very personal considerations. Born in Italy and growing up in the country's Southern province, I am one of those many Italian students who, after a first stop in Milan, Italy's most internationalised city, have decided to continue their trajectory of academic migration abroad. They call us '*cervelli in fuga*', both perpetrators and victims of the country's endemic brain drain. Yet, to consider this phenomenon in purely national terms would be misleading, as I am also an outright child of the much broader university restructuring brought about by the so-called Bologna process with its increased push towards internationalisation and interdisciplinarity (cf., e.g., Moscovitz and Zahavi 2020). An in-depth discussion of these complex phenomena would go well beyond the aims and scope of the present discussion, yet it helps to contextualise my personal encounter with the Chinese Other. As a European social scientist with strong Italian roots yet completely socialised in the German academic environment, I have often felt like a stranger in my country of origin, where the debates at the core of my research interests—among others, in between critical political economy and critical discourse analysis—have long been notably underrepresented. Struggling between Italian origins, German experiences, and Anglophone outreach, China had never represented an option in my plans. And yet it arrived and changed everything. How did it happen?

With the benefit of hindsight, my own China-bound experience of academic migration after the PhD would perfectly corroborate the findings of Braun Střelcová et al. (2022, 23) in terms of motivations—'driven by [my] expected comparative advantage of both gaining new economic and symbolic capital and valorizing [my] existing capital, such as social and cultural capital'—and job satisfaction—being satisfied yet experiencing some main differences between natural and social science domain, with the latter being more disadvantaged in gaining economic and social capital. An interviewee in the study quoted above perfectly summarises the key prerequisite behind this overall match of expectations and reality, featuring my China experience, too. As she comments, '[i]f you're a strong, well-experienced international, you can do it. If you're a European coming here with European ideals, it is hard. You need to have done some substantial experience or work outside of Europe to survive here well. It is not about visiting, but it's about living, and it's not about surviving, but it's about thriving, and in order to do that, you need a different outlook. Things will not work the way they work in Europe. *You need to adapt and observe.*' (Interviewee 16 quoted in Braun Střelcová et al. 2022, 18, emphasis added).

To observe. No other action could better condense the core of my encounter with the Chinese Other. What would you do if you happened to be in a context in which you could not understand what you were told? If you cannot read what is written everywhere? If you cannot interact with people by relying on (some kind of) common language? In such an 'authentic Chinese' city like Wuhan, in which Anglophone interaction in the daily routine of a researcher, teacher, and mum is not straightforward at all, relying on technology and translation apps is, of course, a viable and practical solution; however, it will never cover the entirety of that experience. So, what happens if you are in a situation that you do not feel is under your (language) control? Well, *you observe.* With all your senses. Especially hearing, sight, and smell are activated and challenged in a completely different and unprecedented way compared to situations that you can otherwise manage through language. To put it bluntly, the encounter with the Other will always be more creative than what we might imagine. Through observation, multi-sensorial observation, I could experience how all senses converge into language, leading to completely unexpected implications that no quantitative or even qualitative research will ever be able to grasp in their complexity.

In my case, as a European social scientist starting her academic path from applied linguistics and progressively moving to political theory and political economy as inherently related to my original focus on language and discourse, the real-life encounter with China has been unexpectedly rich in implications. Back in 2018, my international and interdisciplinary expertise had been key to being appointed as a Lecturer at the Huazhong University of Science and Technology. Since then, my teaching and research tasks at the School of Philosophy have consolidated my inclination to act as a *bridge* between my European background and the new Chinese environment and interlocutors—students, colleagues—in several ways. This is the origin of some long-standing courses of mine, such as 'Professional German for Philosophy Students' and 'Introduction to Critical Discourse Studies' for undergraduates and 'Philosophy of Practice: Introduction to Antonio Gramsci' for graduate students. Yet, what I could not predict and, again, what could hardly be grasped in any survey or qualitative research about academic migration, is how powerfully China itself entered my own research agenda. China—the very Other that, besides limited Anglophone interactions at the workplace, I was trying to grasp through images, intuitions, and perceptions—became the subject/object I felt the urgency to engage with.

Two key passages marked this growth process. First, back in December 2019, the countless hours spent reading Antonio Gramsci's writings *and observing people* on my way to the department—in the bus, in the metro—generated the intuition that would provoke a U-turn in my professional path: the idea that Gramsci's textual legacy might offer a unique lens to approach China in its complexity. The scientific elaboration of this intuition eventually ushered in a publication for 'Antipode' promoting a diachronic and philologically sound approach to the potential of a Gramscian perspective in analysing China's peculiar trajectory (Caterina 2021). In the study, I argued that the whole *constellation* of concepts emerging from Gramsci's 'Prison Notebooks'—'intellectuals', 'common sense', 'integral state', 'hegemony', and 'passive revolution'—can be a precious tool in the hands of China scholars interested in exploring several interconnected levels of analysis, such as (i) the constructed nature of Chinese hegemony as the inherently dialectical interplay of consensual and coercive elements; (ii) contested boundary shifts between political and civil society on the terrain of China's integral state; (iii) adopted—and successful—strategies of passive revolution especially within key U-turns in the history of the Chinese Communist Party; (iv) the materiality of Chinese hegemony as grounded in its continuously evolving mode of production; (v) its processual and contested nature across different times and scales; and (vi) key links between Chinese intellectuals and the production of common sense. Suddenly, I was not simply living *in* China anymore; I was *living China* and allowing the Chinese Other—through its very presence—to leave new traces in my vision of the world. In other words, some sort of Gramsci-China encounter was taking place in my daily routine as a teacher and researcher. We are all philosophers, Gramsci argues, as we all feature a vision of the world, a conception of life that relates to a certain ethical attitude. Most importantly, philosophy and politics stay in a continuous dialectical relationship with one another. Every human establishes a unique mental connection to one's own social environment (philosophy) and, on this basis, every human necessarily intervenes in this social environment by transforming it, whereby such transformation is an irreducibly *political* act (Gramsci 1971, 323). Consequently, 'the real philosopher is, and cannot be other than, the politician, the active man who modifies the environment, understanding by environment the *ensemble* of relations which each of us enters to take part in' (Gramsci 1971, 352).

The outbreak of the COVID-19 pandemic boosted this process in which philosophy and politics merge in unforeseen ways, letting China forcefully enter my research agenda again. Few elements, which we now even take for granted or do not remember anymore in their original disrupting and dramatic novelty, may suffice to explain why. Back in January 2020, after the first cases of unknown pneumonia had already been officially reported in Wuhan, a frantic climax of events—from the assessment of human-to-human transmission of the virus, up to the historically unprecedented lockdowns in Wuhan's Hubei region and beyond—eventually managed to reach the headlines of the European press. Initially, European countries looked at such developments as a still external phenomenon. However, swift shifts in the global geography of contagion soon subverted any initial reassuring outlook on a 'safe Europe' as opposed to an 'infected China'. In this context, unexpected parallels between China's Hubei and Italy's Lombardy immediately stood out. Echoing Wuhan's dramatic rise in prominence, Lombardy similarly emerged as a focal point, both regions experiencing transformative moments that shaped their respective trajectories (see Horowitz 2020). These parallels became evident not only in the rapid development and international attention each region received but also in the profound impact on local communities and broader sociopolitical dynamics. The shared experiences of Hubei and Lombardy, marked by sudden change and adaptation, underscore the significance of closely observing and reflecting on such cross-cultural encounters.

In a few weeks, the SARS-CoV-2 contagion dynamics started to unleash controversial processes of identification/othering, solidarity/rejection, inclusion/rejection in Italy-China as well as in intra-European relations that, to date, have only partially been investigated (cf., e.g., Miyake 2021; Pofi and Leung 2022). Who was suddenly 'the Other'? Whom should we be afraid of? How would the pandemic have impacted how we perceive, represent, address, and engage with China? Caught between China and Italy, I could not stop asking these questions. On the one hand, I had left Wuhan on January 16, being on holiday for the Chinese New Year and thus travelling on a fully voluntary basis, just a few days before Hubei's historically unprecedented lockdown would have forever changed the world as we used to know it. On the other hand, I happened to be in Italy, Europe's *bleak heart* of the pandemic, in early 2020. I thus had the privilege to retain my post in China, teaching online at night to reach my Chinese students spread across their cities and villages of origin, yet I was physically in Europe, devouring press reactions to make sense of what was going on. This way, everyday life and changing global circumstances acted as a powerful catalyst on my own research agenda, allowing me a unique entry point in

China debates—in their features, limitations, and the related need to tackle their effects well beyond academic circles.

In Lieu of a Conclusion: Complexity as Resistance

Faced with the capillary, ubiquitous presence of thin China references, I have argued that addressing China's rise as a global actor—such as in the highly strategic sector of S&T—urges a broader self-reflexive outlook on the way *we* perceive, represent, and address the Chinese Other. From a European perspective, I have pointed out key limitations in academic and public China debates—including their monolithic and dichotomous framing, as well as a persistent Anglophone linguistic and cultural bias. Crucially, these limitations are not merely epistemic shortcomings: they are shaped by, and in turn reinforce, underlying political differences between China and European liberal-democratic systems, particularly regarding governance, and the relationship between state, science, and society. As such, political divergence operates as a mediating lens through which China is interpreted—often simplifying complexity while heightening perceptions of systemic contrast and incompatibility.

These limitations should alert us all the more in an increasingly heated geopolitical context, in which such politically inflected framings consolidate and intensify. In such a context, 'we are gradually but decisively slipping from an interpretative paradigm on the "China case" hinging on the question of the "Chinese model" (positive or negative, effective, or ineffective, peaceful or aggressive, etc. [...]) to an interpretative paradigm that I would define as "China as an ideal enemy"' (Samarani 2020, own translation). This shift is itself inseparable from deepening political differences, which not only structure international relations but also reshape epistemic boundaries—what counts as legitimate knowledge, trustworthy collaboration, or acceptable risk. Importantly, in a vicious circle, these heightened political tensions feed into academic debates that, in turn, resonate within civil society, further amplifying polarisation.

The German case is particularly striking in this respect, as it underscores the importance of moving beyond Anglophone analytical frames. In early 2022, a particularly intense controversy unfolded within German China Studies concerning the political and financial entanglements of academic institutions with China. In March 2022, a series of exchanges published in the *Frankfurter Allgemeine Zeitung* (F.A.Z.) between Alpermann and Schubert (2022) followed by a counter-response by Andreas Fulda et al. (2022) ignited a far-reaching debate about the boundaries of academic autonomy, the influence of Chinese state actors, and the transparency of research funding. These concerns were not simply technical or institutional; they were deeply mediated by broader political differences, especially contrasting assumptions about the role of the state in steering research agendas and the legitimacy of such involvement. Just 2 months later, in May 2022, these issues were further amplified by the China Science Investigation, which drew attention to the blurred lines between civilian and military research and the dual-use potential of transnational scientific cooperation (Follow the Money 2022). Taken together, these developments marked a turning point in Germany's academic engagement with China. They illustrate how scholarly debates about research ethics, collaboration and autonomy are increasingly entangled with geopolitical frictions—and, more specifically, how political differences function as a crucial mediating and complexifying factor, shaping both the substance of these debates and the interpretative frameworks through which China itself is understood.

Thus, how to bring 'the human story' (Zhang et al. 2022, 46) back into this exponentially more and more polarised picture? Drawing on my personal experience, I have pointed out how European scientists directly involved in China-bound academic migration, despite representing a necessarily limited group, might play a crucial role as agents for change in advancing a complexity turn in Europe-China relations. Biographical and academic paths, as I have sketched out in my specific case, can merge in unforeseen ways to urge awareness of today's centrality of the 'China Question' (Pavlićević and Talmacs 2022) in our respective political and civil societies. To say it with Gramsci, philosophy, our vision of the world, and politics, our impact on that world, remain in an indissoluble dialectical relation. Contemporary geopolitical challenges, from the Covid-19 pandemic and the Russian invasion of Ukraine to the ongoing war in Gaza, profoundly shape and test this interplay, compelling scholars and citizens alike to reconsider the ethical and political foundations of global engagement. Against this background, I contend, the China Question cannot be (further) delayed for at least three main reasons.

First, the China Question *interrogates* us. 'The issue—Vukovich highlights—is what one says and how one approaches "China", by way of what previously accumulated "knowledge" and one's relationship to that area in the centre of Asia that [...] exists as a political, economic, cultural, and not merely conceptual entity' (Vukovich 2010, 163–164). Thus, what is the 'China' we refer to and with which implications? There is a dire need in scientific *and* public debate to raise awareness of the *contingency* of this question, and thus acknowledge the urgency to take *geo-historical, political, economic, and cultural* factors alike into account in any attempt at providing—however incomplete—answers. To be sure, asking the China Question might leave us with even more interrogatives than those we had in the beginning, as 'Project China'—a documentary on the ineluctable complexity of the Chinese world and of our encounter with its manifold space/time contradictions—powerfully reminds us (Giacchi and Saglia 2020).

Second, the China Question *concerns* us. It is no exclusive domain of Sinologists or foreign policy experts. Contrary to any simplistic dichotomic vision of 'China against the rest', the China question rather concerns us as citizens because of our inherent manifold interconnections with 'Global China' (Franceschini and Loubere 2022). 'Othering' China instead of restoring complexity, relationality, and intersubjectivity is a short-sighted and flawed strategy, which 'on the one hand, [...] obscure[s] the material unity and close interconnectedness of the single [...] omnivorous system/world in which we live, global capitalism, of which China is not only an integral part but also a fundamental driving force. On the other hand, [it] also obscure[s] the infra-cultural diversity, i.e., the plurality of concrete subjects, women and men in the flesh, who inhabit *tianxia*, and the substantial differences in their situations and conditions, material and spiritual, of gender and class' (Crisma 2021, own translation).

Third, the China Question *challenges* us. As scholars, it urges us 'to wander, to contaminate, to derail from the usual tracks' in the direction of a 'hybrid Sinology', as 'China is undoubtedly at the centre of a global evolution that affects no less than all of humanity, and this process is likely to remain only partially intelligible as long as it is viewed from the perch of our fragmented knowledge' (Perini 2021, own translation). Yet, this intellectual challenge is inseparable from the political differences that mediate our engagement with China—differences in governance structures, epistemic cultures, and the relationship between state, knowledge production, and society. These divergences do not simply form a background condition; they actively shape the terms of scholarly inquiry, influencing what kinds of collaboration are achievable, how concepts travel (or fail to travel), and how mutual intelligibility is negotiated or foreclosed. A 'hybrid Sinology', therefore, must also be attentive to how such political differences complicate understanding, requiring not only interdisciplinarity but also a reflexive awareness of the asymmetries and frictions embedded in cross-system exchange.

At the same time, the China Question challenges us as human beings, reminding us of the need to reach out to individuals and restore the centrality of dialogue and one-to-one exchange—even, and especially, in contexts where political divergence risks hardening into epistemic distance or moral closure. In the scientific field, the experience of the 'BioGovernance Commons' discussion group could inspire several similar initiatives across diverse research areas. As the initiators of the group remark, '[w]hile cooperative dialogue with China can achieve lofty geopolitical goals, it's important to also pay attention to the individuals involved. That is to say, instead of denigrating China's rise as an achievement of an autocratic machine, scientists need to recognise the human story in the organization and delivery of those achievements, grasping both the societal reasoning and the social ambivalences involved' (Zhang et al. 2022, 48). Precisely in a context marked by deepening political differences, such initiatives highlight how interpersonal engagement can function as a crucial mediating layer—one that does not dissolve structural tensions but helps to navigate and address them.

The scope of this argument, I contend, reaches well beyond scientific relationships. There is, namely, a dire need 'for different ways of narrating China in the public space: ways that are not stereotyped, not constrained by the inertia of clichés, that do not reduce it to an authoritarian fetish, an alien monolith, but that strive to grasp its irreducible complexity, without demonizing or mythologizing it' (Crisma 2021, own translation). To be sure, such a complexity turn is ill-suited to meet the pressing demand we are all exposed to in a frantic attempt to get political, economic, and cultural ready-to-go solutions. To dare such a turn is thus per se an act of resistance: it is in our hands to make it infectious in scientific China encounters and beyond.

Acknowledgement This chapter was developed within the framework of the research project 'Italy–China: A New Common Sense? Shifting Relations in Pandemic Times' (project number: 2021WKYXQN057), supported by 'the Fundamental Research Funds for the Central Universities'.

References

Alpermann, Björn, and Gunter Schubert. 2022. *Gegen das moralische Kreuzrittertum*. Frankfurter Allgemeine Zeitung, March 3.

Bauder, Harald. 2020. International mobility and social capital in the academic field. *Minerva* 58 (3): 367–387. https://doi.org/10.1007/s11024-020-09401-w.

Braun Střelcová, Andrea, Yuzhuo Cai, and Wei Shen. 2022. The experience of European researchers in China: A comparative capital advantage perspective. *Journal of the Knowledge Economy*1–29. https://doi.org/10.1007/s13132-022-00982-3.

Brands, Hal, and John Lewis Gaddis. 2021. The New Cold War: America, China, and the Echoes of History. Foreign Affairs 100, no. 6 (November/December): 10–20.

Cai, Li, and Christine Hall. 2016. Motivations, expectations, and experiences of expatriate academic staff on an international branch campus in China. *Journal of Studies in International Education* 20 (3): 207–222. https://doi.org/10.1177/1028315315623055.

Caterina, Daniela. 2021. Gramsci in China: Past, present, and future of a still open encounter. *Antipode* 53 (5): 1357–1376. https://doi.org/10.1111/anti.12731.

Crisma, Amina. 2021. Quale ruolo per la centralità della Cina nello spazio pubblico? Proposte, esperienze, fragore di silenzi e clamore di Grandi Narrazioni. *Sinosfere*, March 14.

European Commission. 2019. *Joint communication to the European Parliament, the European council and the council*. EU-China: A strategic outlook.

Follow The Money. 2022. *China science investigation*. Available from https://www.ftm.eu/chinascienceinvestigation

Franceschini, Ivan, and Nicholas Loubere. 2022. *Global China as method*. Cambridge: Cambridge University Press.

Fulda, Andreas, Mareike Ohlberg, David Missal, Fabian Horst, and Sascha Klotzbücher. 2022. Grenzenlos kompromissbereit? *Frankfurter Allgemeine Zeitung*, March 16.

Giacchi, Elisabetta, and Thomas Saglia. 2020. *Project China*. Available from https://giacchielisabetta.wixsite.com/projectchina/il-documentario

Gramsci, Antonio. 1971. *Selections from the prison notebooks*. New York: International Publishers.

Groenewegen-Lau, Jeroen. 2024. Whole-of-nation innovation: Does China's socialist system give it an edge in science and technology? *Current History* 123 (Supplement 1): 19–28. https://doi.org/10.1525/curh.2024.123.S1.19.

Heimer, Maria, and Stig Thøgersen, eds. 2006. *Doing fieldwork in China*. Copenhagen: Nordic Institute of Asian Studies (NIAS) Press.

Horowitz, Jason. 2020. *We take the dead from morning till night*. New York Times, March 27.

Iriye, Akira. 1967. *Across the Pacific: An inner history of American-East Asian relations*. New York: Harcourt, Brace & World Inc.

Isaacs, Harold R. 1958. *Scratches on our minds: American images of China and India*. New York: Harper Torchbooks.

Jones, Lee, and Shahar Hameiri. 2021. *Fractured China: How state transformation is shaping China's rise*. Cambridge University Press.

Mackerras, Colin. 1989. *Western images of China*. Oxford: Oxford University Press.

Mackerras, Colin. 2013. *China in my eyes: Western images of China since 1949*. Beijing: Renmin University of China Press.

Mallapaty, Smriti. 2022. What Xi Jinping's third term means for science. *Nature* 611:20–21. https://doi.org/10.1038/d41586-022-03414-z.

Meinhof, Marius, Junchen Yan, and Lili Zhu. 2017. Postcolonialism and China: Some introductory remarks. InterDisciplines. *Journal of History and Sociology* 8 (1): https://doi.org/10.4119/UNIBI/indi-v8-i1-166.

Miyake, Toshio. 2021. 'Cin ciun cian' (ching chong): Yellowness and neo-orientalism in Italy at the time of COVID-19. *Philosophy & Social Criticism* 47 (4): 486–511. https://doi.org/10.1177/01914537211011719.

Moscovitz, Hannah, and Hila Zahavi. 2020. *The Bologna process and its global strategy: Motivations and external responses*. Routledge.

National Bureau of Statistics of China (2024) China's Innovation Index in 2023. November 7. Retrieved from https://www.stats.gov.cn/english/PressRelease/202411/t20241107_1957293.html

Pan, Chengxin. 2012. *Knowledge, desire and power in global politics: Western representations of China's rise*. Cheltenham: Edward Elgar Publishing.

Pavlićević, Dragan, and Nicole Talmacs. 2022. Answering the "China question": Local responses to global China. In *The China question: Contestations and adaptations*. Singapore: Springer Nature Singapore.

Perini, Gaia. 2021. Giovanni Arrighi a Pechino, Elon Musk a Shanghai. Alla ricerca di una sinologia "ibrida". *Sinosfere*, February 28.

Pofi, Maria Paola, and Wing-Fai Leung. 2022. Responses to health risk and suffering: 'China'in the Italian media discourses during the early stage of the Covid-19 pandemic. *Media, Culture & Society* 44 (4): 820–836. https://doi.org/10.1177/01634437211053770.

Samarani, Guido. 2020. Noi e la Cina. Né filocinesi, né anti-cinesi. *Sinosfere*, November 14.

Seib, Philip, and Shawn Powers. 2010. *China in the news: A comparative analysis of the China coverage of BBC world service, CNN international, and deutsche Welle*. The USC Center on Public Diplomacy.

Sun, Tara Qian, Yun Liu, Xiaolan Fu, and Juan Rogers. 2025. Global innovation cooperation and governance: Evidence from China. *Science and Public Policy* 52 (3): 436–437. https://doi.org/10.1093/scipol/scae089.

Turner, Oliver. 2014. *American images of China*. New York: Routledge.

Vukovich, Daniel. 2010. China in theory: The orientalist production of knowledge in the global economy. *Cultural Critique* 76:148–172.

Wang, Bingyu, and Jingfu Chen. 2021. Emotions and migration aspirations: Western scholars in China and the navigation of aspirational possibilities. *Journal of Ethnic and Migration Studies* 47 (15): 3437–3454. https://doi.org/10.1080/1369183X.2020.1764841.

Xiao, Siwen, and Yaosheng Xu. 2024. *(Re) centralization: How China is balancing central and local power in science, technology, and innovation*. Working paper, UCIGCC.

Yan, Wenjie. 1998. A structural analysis of the changing image of China in the New York times from 1949 through 1988. *Quality and Quantity* 32 (1): 47–62. https://doi.org/10.1023/A:1004245831853.

Zhang, Joy Yueyue, Sonia Ben Ouagrham-Gormley, M. Kathleen, and Vogel. 2022. Creating common ground with Chinese researchers. *Issues in Science and Technology* 38 (4): 45–48.

Zhang, Li. 2010. The rise of China: Media perception and implications for international politics. *Journal of Contemporary China* 19 (64): 233–254. https://doi.org/10.1080/10670560903444199.

Zweig, David, and Huiyao Wang. 2013. Can China bring back the best? The Communist Party organizes China's search for talent. *The China Quarterly* 215:590–615. https://doi.org/10.1017/S0305741013000751.

Daniela Caterina is Associate Professor at the School of Philosophy at Huazhong University of Science and Technology (China), where she has been based since 2018, following her appointment as Lecturer. She holds a PhD in Political Science from the University of Hamburg (*summa cum laude*), as well as an MA in European Studies from the same institution. Her doctoral research examined the struggle for hegemony in Italy's crisis management, with a particular focus on the 2012 labour market reform, combining insights from cultural political economy and critical policy analysis. Alongside her current position, she has held visiting and teaching appointments at several European institutions, including the University of Vienna, the Catholic University of Milan, Bielefeld University of Applied Sciences, and the University of Campania 'Luigi Vanvitelli'.
Her research lies at the intersection of political theory, international political economy, and critical discourse studies, with a regional focus on Europe and China. She has published extensively on themes such as technocracy, neo-authoritarianism, economic imaginaries, and the transformation of political space, often drawing on Gramscian perspectives and cultural political economy. She is the author of the monograph *Struggles for Hegemony in Italy's Crisis Management* (Springer, 2019), and her forthcoming book *Discourses of China as Other: Intellectual Production and Political Economy* (Routledge, 2027) further develops her work on discourse, power, and global political economy. Her work contributes to advancing critical-qualitative approaches in political analysis, with a particular emphasis on the role of ideas, discourse, and intellectual production in shaping contemporary political and economic transformations.

Chapter 4
What Does "Adaptive Governance" Mean in the Chinese Biobank Context?

Haidan Chen

Abstract This chapter examines what "adaptive governance" means in the specific context of Chinese biobanks. Drawing on over 15 years of fieldwork and sustained involvement in international initiatives such as BIONET and the BioGovernance Commons, it traces the rapid expansion of biobanking infrastructures in China and the evolution of a distinctive regulatory regime centered on human genetic resources. The analysis shows how Chinese biobank governance has been shaped by concerns over bio-sovereignty, national security, and the prevention of bio-piracy, resulting in a predominantly top-down, state-led regulatory framework. It critically interrogates notions of adaptive governance, commonly associated with flexibility, social learning, and multi-stakeholder engagement. It argues that while elements of adaptive governance have emerged in China, they remain constrained. The chapter calls for a dual adaptation: not only institutional mechanisms that enhance transparency, accountability, inclusiveness, and long-term sustainability, but also a cultural and cognitive shift in regulatory mindsets. Recognizing bottom-up initiatives, public participation, and transnational dialogue is essential if adaptive governance is to support China's evolving role in global biomedical research.

Keywords Biobanks · Adaptive governance · China · Human genetic resources · Bio-sovereignty · Bio-piracy

Introduction

Since September 2021, I have been actively participating in the monthly online meetings of BioGovernance Commons, a transnational initiative jointly convened by Joy Y. Zhang, Kathleen Vogel, Sonia Ben Ouagrham-Gormley, and Ruipeng Lei

H. Chen (✉)
School of Health Humanities, Peking University, Beijing, China
e-mail: chenhaidan@bjmu.edu.cn

J. Y. Zhang et al. (eds.), *BioGovernance Commons and China's Role in Global Risk Governance*, Emerging Globalities and Civilizational Perspectives,
https://doi.org/10.1007/978-3-032-32198-5_4

for "sharing perspectives on shared challenges." This platform serves as a bridge connecting scholars from China and the West to discuss various areas of life science governance from cross-cultural and interdisciplinary perspectives. As a scholar based in a Chinese university whose research has focused on the governance of emerging medical technologies since 2006, I was often expected to have a good grasp of the Chinese situation and was therefore frequently asked to share insights into specific areas of life science governance here in China. Such exchanges revealed a striking reality: outsiders are eager to understand life science governance in China, a non-Western rising scientific power, yet the country's actual situation remains opaque to much of the world.

The lack of clarity became especially evident in the wake of the "gene-edited babies" incident in November 2018 (Meyer 2022). International colleagues were curious about how and why this could happen in China, what exactly Jiankui He had been doing, and what consequences he would face. Through ongoing discussions, including the invitation extended to Jiankui He to join a BioGovernance Commons event after his release from prison in 2023, participants sought to grapple with two central issues: the governance gaps highlighted by his case and the broader responsibilities of global scientific communities. These moments not only revealed the frustrations surrounding China's science but also underscored the need for deeper engagement with how life science governance is enacted and negotiated across different cultural and political contexts.

It was against this backdrop of cross-cultural dialogue and critical reflection that I turned to the issue of biobank governance in China, a field I have been engaged with since 2009. My interest in this area first emerged when I was a junior research fellow in the international collaborative project BIONET (Ethical Governance of Biological and Biomedical Research: Chinese-European Co-operation, 2006–2009). Apart from the opening and closing conferences in London, the five other BIONET workshops were held in different Chinese cities, bringing together Chinese and European partners, along with other invited participants from China and abroad, to foster mutual understanding of research cultures and life science governance in China and the West.

On 27–30 April 2009, I attended the final BIONET workshop on "biobanks and genomic research," hosted by the Beijing Genomics Institute (BGI) in Shenzhen. This was my first encounter with biobank-related discussions. This workshop brought together about 60 scientists, social scientists, ethicists, clinicians, and biobankers from China and the West. At that time, only a handful of biobanks had been established in China, and while several researchers were planning new ones, conversations on biobank governance among Chinese biobankers and scholars were still scarce. In contrast, many biobanks were already operating in Europe and North America, where governance issues were a prominent focus. At the workshop, I met two colleagues from Shanghai Clinical Research Center (SCRC), who aimed to learn from international experience as SCRC prepared to launch its own biobank. Over the past decade, SCRC, alongside other Chinese medical institutions, has successfully set up biobanks and worked to address ethical and legal challenges, including informed consent, data sharing, privacy, and confidentiality. At the national level, China has increasingly prioritized biobank development in policy documents and strengthened the protection of human genetic resources through new regulatory measures.

Since 2009, I have examined biobank governance in China through qualitative methods, including case studies, in-depth interviews, participatory observation, and document analysis. Building on these experiences, this chapter reviews the evolution of China's biobanking sector and the accompanying regulatory debates. It interrogates the persistent self-positioning of China as a developing country in its policy documents and contrasts this lingering Third-World mentality with its simultaneous pursuit of global leadership. This is not to deny China's colonial past or its concerns about bio-piracy and bio-sovereignty. Rather, it underscores the dual form of "adaptiveness" needed for China to sustain its scientific ascent: on the one hand, the conventional mode of adaptive governance based on decentralized, multi-stakeholder deliberation; on the other, an adaptation of the national psyche to its growing power and responsibilities.

The Emergence of Biobanks and Biobank Networks

Biobanks collect biological samples and associated data and are widely regarded as critical infrastructures for accelerating the translation of basic medical research into clinically applicable diagnostics and therapeutics, ultimately contributing to the development of personalized medicine. Modern biobanks range from disease-specific biobanks to large-scale population-based cohort studies, and are funded by public institutions, pharmaceutical companies, disease advocacy organizations, and charitable foundations. Biobanking resources support academic research, clinical studies, biotechnology development, and even forensic applications. In the post-genomic era, a growing number of biobanks have been established at local, national, and international levels to advance health research (O'Donoghue et al. 2022).

Despite the enthusiasm surrounding biobanks, the establishment and maintenance of both physical and virtual biobanks are highly resource-intensive and costly. Networking and collaboration among existing biobanks can help reduce demands on personnel, materials, and financial resources; enhance the utilization of stored samples and data; and facilitate the research and development of diagnostic and therapeutic products (Coppola et al. 2019). Consequently, fostering collaboration and establishing biobank networks have become an increasingly prominent trend in the field. Several regional, national, and international infrastructures, such as the Biobanking and Biomolecular Resources Research Infrastructure (BBMRI) in Europe, the National Center Biobank Network in Japan, and the International Agency for Research on Cancer Biobank, have been created to integrate individual biobanks into larger collaborative networks.

A central feature of biobank networks is sharing samples and data while ensuring the interoperability of these resources within ethically and legally acceptable frameworks (Kiehntopf and Krawczak 2011). Although large-scale international collaborations, such as the Human Genome Project and the International HapMap Project, have promoted open data access, many biobanks remain reluctant to share samples and data. This reluctance stems from the ethical and legal concerns, including privacy, confidentiality, commercialization, and ownership (Boeckhout and Mayrhofer 2019). Academic competition, disputes over co-authorship, and concerns about fairness further affect biobankers' willingness to participate in biobank networks and share resources (Devriendt et al. 2021; Colledge and Elger 2015). Furthermore, the absence of standardized procedures for sample handling and quality management continues to pose significant obstacles to biobank integration. As a result, many biobanks and networks are still struggling to address these challenges (Watson et al. 2011; Argudo-Portal and Domènech 2020).

A Need for Adaptive Governance

These challenges have prompted growing scholarly attention to how biobanks should be governed. Much of the existing literature has approached biobank governance through the lens of the ethical, legal, and social implications (ELSI) of biobanking (e.g., Boeckhout and Mayrhofer 2019; Gottweis and Petersen 2008). More recently, risk assessment and management have been proposed as complementary strategies to address the complexities and uncertainties inherent in biobanking practices. For example, Rychnovská identified three major types of risks: data security (protection against unauthorized access), privacy (protection of sensitive information), and data misuse (protection against societal harm), and called for anticipatory governance in biobanking, emphasizing the need to foresee and mitigate potential risks before they materialize.

Building on this perspective, Akyüz et al. (2021) proposed a broader typology of biobank-related risks within an adaptive risk governance framework. This approach extends beyond anticipation by emphasizing flexibility, ongoing learning, and stakeholder engagement in responding to evolving challenges. They distinguish between risks to biobanks and risks to participants. Biobank-related risks include economic risks (e.g., loss of sustainable funding), technical risks (such as equipment or freezer failures resulting in biospecimen loss), occupational risks to employees handling biological samples, participation risks arising from declining public trust, and research community risks, including scientific misconduct or misuse of biobanking resources that may erode societal confidence. Participant-related risks include physical harm, informational/privacy risks, and risks associated with the disclosure of individual or incidental findings. The long-term establishment and maintenance of biobanking infrastructures are therefore inherently complex and fraught with governance challenges. In this context, adaptive governance has emerged as a promising framework for navigating uncertainty and institutional fragility.

Adaptive governance rests on the premise that, under conditions of uncertainty, limited evidence, and evolving socio-technical dynamics, a governance system should be capable of adjusting to the evolving phenomena they seek to regulate. As Vayena and Blasimme (2021) argue, key attributes of adaptive governance include multi-stakeholder participation in governance design, collaboration between regulators and regulated actors, incremental and evidence-based policy development, and openness to uncertainty through processes of social learning. Social learning is particularly crucial, as it enables diverse actors to interpret emerging risks, reassess normative assumptions, and refine governance responses over time. Individuals, agencies, organizations, and institutions contribute heterogeneous forms of expertise and experience, interacting across levels and scales to inform policy adaptation. In this way, adaptive governance aims to balance top-down regulatory coordination with bottom-up solutions, reconciling centralized authority with decentralized flexibility (Rijke et al. 2012; Vayena and Blasimme 2021).

China provides a particularly illuminating case in this regard. Like many other countries, it has invested substantial resources in developing biobanks to address national public health priorities, respond to the pressures of a rapidly aging population, promote biomedical innovation, advance translational medicine, and strengthen global competitiveness in the life sciences. A distinctive feature of the Chinese regulatory approach, however, is its longstanding emphasis on safeguarding human genetic resources, an objective embedded in national regulations since the late 1990s (Sung 2009). In the sections that follow, I first examine the development of biobanks in China, then trace the evolution of the national regulatory framework and its implications for biobank governance, and finally explore what "adaptive governance" entails within the specific institutional and political context of Chinese biobanking.

Biobank Expansion and Network Formation in China

Biobanking is both a longstanding practice and an emerging scientific discipline. Pathologists have collected and stored biological samples in hospitals for decades, yet contemporary biobanking integrates advances in information technology, molecular medicine, and data science to facilitate biomedical research (Morente et al. 2008). In China, hospitals began to collect biological samples for internal research purposes as early as the 1970s. One of the earliest formal biobanks, the Chinese Different Ethnic Groups' Immortalized Cell Line Bank, was established in 1994 by the Institute of Medical Biology at the Chinese Academy of Medical Sciences in Kunming, Yunnan Province (Chu et al. 1998). Following the completion of the Human Genome Project, the systematic collection of biological samples and associated data gained increasing recognition as a foundation for genomic research and translational medicine. Although the development of biobanks in China progressed relatively slowly during the 2000s, significant expansion occurred during the 12th (2011–2015) and 13th (2016–2020) Five-Year Plans for National Economic and Social Development.

During the 2000s, only a limited number of large-scale biobanks were established. The China Kadoorie Biobank (CKB), founded in 2004 with initial funding from the Kadoorie Charitable Foundation, represents a major international collaboration involving the University of Oxford Clinical Trial Service Unit and Epidemiological Studies Unit, the Chinese Center for Disease Control and Prevention, and the Chinese Academy of Medical Sciences. As a prospective cohort study of over 0.5 million Chinese participants, the CKB investigates lifestyle, environmental, and genetic determinants of common diseases (Chen 2013). Similarly, the Taizhou Biobank, established in 2007 in Taizhou, aimed to explore environmental and genetic risk factors for prevalent diseases (Wang et al. 2009). The Beijing Biobank of Clinical Resources, launched in 2009 with support from the Beijing Municipal Science and Technology Commission and coordinated by Capital Medical University, focused on major diseases such as diabetes, cardiovascular diseases, and cancer through collaboration among eleven leading hospitals in Beijing (Li et al. 2017).

The 12th Five-Year Plan (2011–2015) marked a turning point in state-level recognition of biobanks as strategic scientific infrastructure. The 12th Five-Year Plan for Biotechnology Development (2011) called for the construction of major scientific and technological infrastructures, including large-scale biobanks, to promote resource sharing (Ministry of Science and Technology 2011). Similarly, the 12th Five-Year Plan for the Development of Strategic Emerging Industries (2012) identified the establishment of national gene banks and biobanks as a priority for advancing the biomedical industry (State Council 2012). As a result, the number of biobanks expanded significantly during this period.

The 13th Five-Year Plan (2016–2020) further underscored the role of biobanks as strategic assets essential for national drug development and public health, while placing greater emphasis on integration, standardization, and coordination. The National Science and Technology Innovation Plan proposed the establishment of a national strategic biological resource database and information service platform to ensure sustainable development of the biotechnology sector (State Council 2016). Subsequent policy initiatives, including the 13th Five-Year Plan for Health and Medical Science and Technology Innovation and the Optimization and Integration of National Science and Technology Innovation Bases, aimed to establish nationally interconnected large-scale biobank platforms and to promote standardization and integration of biobanking resources to support the prevention and treatment of major diseases (Ministry of Science and Technology et al. 2017a, b)

Influenced in part by the U.S. Precision Medicine Initiative (later renamed the All of Us Research Program), China incorporated precision medicine into its national research agenda. In 2016, the Ministry of Science and Technology issued guidelines under the National Key Research and Development Program to support precision medicine research. These initiatives included the establishment of large-scale cohort studies and integrated platforms for storing, utilizing, and sharing precision medicine data. Cohorts were categorized into three types: national population-based cohorts, major disease-based cohorts, and rare disease-based cohorts. The national population-based cohorts were further distributed across seven geographic regions of China: Northeast China, North China, Northwest China, East China, Central China, Southwest China, and South China. Major disease-based cohorts covered conditions including rheumatic diseases, neurological disorders, mental illnesses, and five major cancers. The China Rare Diseases Clinical and Genomic Database aimed to recruit over 50,000 patients across 50 rare diseases. Data generated from these cohorts were centralized for bioinformatics analysis, with the objective of translating findings into clinical decision-making tools and public health strategies.

In practice, most Chinese biobanks are operated by hospitals, hospital departments, or research institutes. Examples include hospital-based biobanks focusing on oncology or hematology, as well as national-level infrastructures such as the China National GeneBank in Shenzhen. To enhance coordination and improve resource utilization, biobank networks have been actively promoted. The Shanghai Biobank Network (SBN), established in 2012 with municipal support, sought to position Shanghai as a hub for biopharmaceutical innovation and clinical research (Gan et al. 2015). The national precision medicine initiative further reinforced efforts to build interconnected regional and national biobank networks.

Despite rapid expansion, structural challenges persist. A national survey of 70 biobank managers found that most disease-based biobanks were established and operated within Grade A tertiary hospitals, yet their geographic distribution was uneven, with a strong concentration in economically developed eastern and northern regions. Blood-derived specimens were the most commonly stored sample type, and the number of stored samples ranged from fewer than 10,000 to more than two million. Although large-scale biobanks reported sharing resources with external institutions, most samples and data remained internally controlled, and overall utilization rates were relatively low (Chen et al. 2021). A separate survey of 63 Chinese biobanks reached similar conclusions, indicating that utilization rates in most institutions were below 20%, with samples predominantly reserved for internal research purposes (Li et al. 2018).

Regulatory Trajectory of China's Biobanks

In China, biological samples and associated data collected and stored in biobanks are legally categorized as human genetic resources and governed under a dedicated regulatory framework. The earliest regulation, the Interim Measures for the Administration of Human Genetic Resources (hereafter the Interim Measures), was promulgated in 1998 by the Ministry of Science and Technology and the Ministry of Health with the approval of the State Council (Ministry of Science and Technology and Ministry of Health 1998). Its stated purpose was "to efficiently protect and rationally utilize human genetic resources in the People's Republic of China, strengthen the research and development of human genes, and promote international cooperation and exchange on the basis of equality and mutual benefits"

(Article 1). Human genetic resources were classified as state scientific or technological secrets (Article 5).

Under the Interim Measures, foreign institutions or individuals seeking to use Chinese human genetic resources for research purposes were required to collaborate with Chinese counterparts, who assumed responsibility for obtaining official approval. The export of important human genetic resources or their provision to foreign entities was placed under strict administrative control (Article 14). To enforce these requirements, the Human Genetic Resources Administration of China (HGRA) was established to review and approve international collaborative projects.

The Interim Measures emerged amid concerns over alleged bio-piracy by foreign researchers. During implementation, however, regulatory loopholes became apparent, as certain institutions and investigators circumvented approval procedures and exported biological samples and associated data without HGRA authorization. In response, the Ministry of Science and Technology, together with other agencies, initiated the drafting of the Regulations on the Administration of Human Genetic Resources (hereafter the Draft Regulations) in 2006. The drafting process involved consultations with government departments, research institutions, enterprises, and experts, and a draft version was released for public comment in 2012. While retaining the core principles of the Interim Measures, the Draft Regulations introduced several new provisions. These included a qualification and licensing system for institutions engaged in the collection and preservation of human genetic resources; strengthened supervision over collection, preservation, research, and utilization activities; clarification of national interests in international cooperation; and more explicit provisions concerning legal liability for violations.

Because the Draft Regulations were not immediately finalized, the Ministry of Science and Technology issued the Administrative Licensing Service Guide for the Review and Approval of the Collection, Preservation, Trade, and Export of Human Genetic Resources (hereafter the Service Guide) as an interim administrative instrument. Effective from October 1, 2015, the Service Guide largely reflected the substantive requirements of the Draft Regulations. It reiterated the principles of mutual benefit in international cooperation while emphasizing the protection of national security, public interests, and public safety (Ministry of Science and Technology 2015).

Following extensive revisions and expert review, the Regulations on the Administration of Human Genetic Resources were formally promulgated in May 2019 and entered into force on July 1, 2019. Under these regulations, foreign organizations, institutions, or individuals must collaborate with Chinese entities and submit joint applications for HGRA approval (Article 22). Applications are subject to multiple substantive conditions, including ensuring no harm to public health, national security, or societal interests; obtaining ethical approval in both jurisdictions; and establishing clearly defined benefit-sharing arrangements. Where genetic data are to be exported, an official exit certificate must be obtained and customs procedures completed. The chapter on legal liability establishes substantial administrative penalties for noncompliance (Ministry of Science and Technology 2019). Across successive regulatory instruments, the protection of human genetic resources from foreign exploitation has remained a central concern. Both the Interim Measures and the 2019 Regulations are primarily structured around principles of bio-sovereignty, national security, and equitable benefit sharing in international collaboration. By contrast, issues such as informed consent and privacy protection receive comparatively limited and less systematic treatment

The sovereignty-oriented framework was further consolidated in 2020 through two major legislative developments. The Biosafety Law of the People's Republic of China (promulgated October 17, 2020; effective April 15, 2021) explicitly affirms state sovereignty over human genetic and biological resources and mandates strengthened supervision over their collection, preservation, utilization, and external transfer (Article 53), with corresponding penalties provided in subsequent provisions (National People's Congress 2020a). Similarly, Amendment XI to the Criminal Law (promulgated December 26, 2020; effective March 1, 2021) introduced criminal liability for the unlawful collection, transportation, or export of human genetic resources in ways that endanger public health or social interests (Article 38) (National People's Congress 2020b).

In practice, however, the regulatory approval process has been criticized by biobank practitioners as lengthy and administratively burdensome. In response to these concerns, the Ministry of Science and Technology issued notices in 2017 and 2020 to streamline and expand the use of simplified procedures. Persistent implementation challenges prompted further consultations, organized by the China National Center for Biotechnology Development, with research institutes and biobanks to develop more detailed implementation rules. These implementing rules were released for public comment in March 2022 and formally adopted on May 26, 2023, entering into force on July 1, 2023 (Ministry of Science and Technology 2023).

In summary, since 1998, China's regulatory trajectory in the biobank domain has been primarily oriented toward safeguarding bio-sovereignty, national security, and equitable terms of international collaboration. Only in recent years has greater emphasis been placed on domestic integration and standardization of biobanking resources, particularly through centralized data-sharing platforms. For example, the 2016 Precision Medicine Research Project mandated the sharing of cohort data, biospecimens, and clinical information with a national big data platform. Likewise, the 2018 Administrative Measures for Scientific Data introduced requirements to standardize data management, enhance security safeguards, and promote controlled openness in support of scientific innovation. Taken together, these developments suggest that China's regulatory framework has prioritized sovereignty and macro-level data governance, while comparatively less systematic attention has been devoted to individual-level concerns such as informed consent, privacy, and confidentiality.

Bottom-Up Ethical and Professional Initiatives

Complementing the top-down regulatory framework, a range of bottom-up initiatives have been advanced by biobank practitioners, professional associations, and ethics review boards. For instance, drawing on international guidelines and norms, the Shanghai Clinical Research Center Ethics Committee issued the Ethical Management Guidelines for the Shanghai Disease-Based Biobank Network (hereafter the Shanghai Guidelines) in 2014 (Chinese version) and 2015 (English version). These guidelines aimed to safeguard participants' interests while standardizing the construction, management, and resource-sharing practices of the Shanghai Disease-Based Biobank Network (Zhu et al. 2015).

The Shanghai Guidelines consist of seven chapters: general principles; informed consent, use of biological samples from individuals lacking consent capacity; privacy and confidentiality; application procedures for samples and data access; intellectual property and resource sharing; and conflict of interest, each offering detailed ethical and operational recommendations. With respect to informed consent, the guidelines endorsed a broad consent model, on the grounds that traditional project-specific consent was impracticable in biobanking contexts. Acknowledging the risks of discrimination and stigmatization resulting from privacy breaches, the guidelines introduced detailed de-identification requirements and mandated that participating institutions establish robust information security systems. As one of the earliest comprehensive ethical frameworks for biobanking in China, the Shanghai Guidelines were adopted by several institutions. In 2020, the committee further issued the "Model for Ethics Review of Human Biobanks" intended as a reference framework for ethics committees nationwide (Shanghai Clinical Research Ethics Committee 2020).

In the early stages of biobank development in China, standardized procedures for quality control, data management, and governance were largely absent. Research institutes and hospitals often prioritize rapid sample collection with limited attention to ELSI considerations. Some biobank managers regarded stored biospecimens as quasi-private assets and were reluctant to share them beyond their institutions. As a result, many biobanks operated in a fragmented, poorly coordinated, and relatively closed manner. Concerns emerged regarding sample quality and long-term usability. Some scientists and regulators warned that numerous collections had already become degraded and unusable, and that biobanks risked turning into inactive or "dead" repositories (Zhang 2018). In this context, biobanks and networks evolved largely through trial and error. Improving standardization and enhancing effective utilization became urgent priorities for both practitioners and policymakers.

To advance professionalization, standard-setting, and international exchange, the Biobank Branch of the China Medical Biotechnology Association (BBCMBA) was established in 2009, approved by the Ministry of Health and the Ministry of Civil Affairs, and coordinated by the National Engineering Center for Biochip in Shanghai. Guided by the principle of "cherishing samples, implementing standards, maximizing utilization, and safeguarding property rights," the BBCMBA actively promoted standardized biobank development. It convened a panel of experts to draft the Standards of Biobanks (Trial), published in the Chinese Medicinal Biotechnology journal in February 2011, and invited feedback for further refinement (BBCMBA 2011).

Despite its efforts to encourage data sharing, the BBCMBA encountered persistent institutional resistance. Most research institutes and hospitals within its network were unwilling to share samples. A senior manager at the National Engineering Center for Biochip in Shanghai observed:

The most difficult task [in building a biobank network] is persuading research institutes to share their samples. Researchers treat their collections as personal assets. Competition in Chinese life sciences is intense. Sample owners fear losing priority in publication if others gain access. Even when co-authorship is offered, disputes frequently arise over first or corresponding authorship, both of which are crucial in research evaluation. Furthermore, property rights remain unresolved; we still lack clear mechanisms to address these issues. (Personal communication, July 6, 2013).

Such concerns continue to influence biobank governance practices.

The BBCMBA also organized annual biobank conferences beginning in 2009 and developed close collaborations with the International Society for Biological and Environmental Repositories (ISBER). It invited ISBER presidents and experienced international biobank professionals to its meetings. After the publication of ISBER's Best Practices in 2012, the BBCMBA coordinated the Chinese translation and offered training programs to disseminate these standards domestically. ISBER, in turn, welcomed Chinese biobankers to its annual conferences and co-hosted its 2019 annual meeting in Shanghai in partnership with the BBCMBA. These international exchanges facilitated the diffusion of professional norms and operational expertise. In parallel, numerous local, regional, and national workshops and conferences provided important platforms for knowledge sharing and professional networking.

Further institutionalization of biobank standardization occurred in June 2015 with the establishment of the National Technical Committee 559 on Biospecimens under the Standardization Administration of China. Comprising more than 60 biobank professionals and bioethicists, the committee assumed responsibility for drafting and revising national standards governing the collection, processing, storage, management, distribution, and application of biological samples. Since its establishment, several key standards have been issued, including the General Requirements of Quality and Competence for Biobanks (2019), the Ethical Requirements of Human Biobanking (2020), the Management Specification for Human Biobanks (2021), and the Fundamental Terminology for Human Biobanks (2021). Although these initiatives have not eliminated all structural challenges, they have substantially strengthened the professional and ethical infrastructure of Chinese biobanking.

Discussion and Conclusion

In its ambition to become a global leader and a scientific powerhouse in biomedicine and life sciences, China has made tremendous efforts, including undertaking wide-ranging reforms and increasing investment to enhance scientific productivity (Kennedy 2019). Building standardized biobanks with high-quality biological samples and associated data is crucial for advancing health research, translational medicine, and the biomedical industry.

Over the past decade, biobanks have developed rapidly in China. The Chinese government has adopted a top-down regulatory approach, treating biobanks as repositories of human genetic resources and promulgating new regulations to strengthen protection in the name of national interests and public health. Any export of human genetic resources for research purposes needs to be reviewed and approved by designated authorities; otherwise, violators face severe penalties. In recent years, national policies have also emphasized the standardization, integration, and sharing of biobanking resources to support health research, disease diagnosis, and treatment. Centralized scientific data management systems have been established to consolidate and coordinate data resources. Overall, the State has sought to maintain firm control over human genetic resources, both domestically and in the context of international collaboration.

At the same time, Chinese biobankers have self-organized into professional networks and actively engaged in social learning from international experiences and knowledge systems. They have initiated a range of activities, such as inviting experienced international biobankers to speak at conferences, organizing and participating in professional meetings, translating the ISBER Best Practices from English into Chinese, providing training courses, and publishing management guidelines and technical specifications for human biobanks. These bottom-up efforts have significantly promoted the standardization and integration of biobanks in China. There has also been interaction between the regulatory actors (e.g., the Ministry of Science and Technology, the China Medical Biotechnology Association, and the HGRA) and regulated actors (e.g., biobankers, hospitals, and research institutes) in both the policy-making and implementation of the Regulations on the Administration of Human Genetic Resources. Despite such interactions, regulators have largely dominated the focus and direction of biobank governance, prioritizing the protection of human genetic resources against concerns such as bio-piracy.

This state dominance is underscored by high-profile enforcement actions taken by the central government. Between 2015 and 2018, the Ministry of Science and Technology issued landmark administrative penalties against major domestic and international players, including BGI and AstraZeneca, for unauthorized data sharing or the illegal export of samples (Wang 2018). These enforcement precedents paved the way for the 2019 Regulations, which significantly increased administrative fines to up to RMB 10 million (or ten times the illegal income) for severe violations. This regulatory regime was further reinforced by Amendment XI to the Criminal Law in

2021, which officially criminalized the illegal collection and exportation of human genetic resources.

The life cycle of biobanks involves numerous individuals, agencies, organizations, and institutions in the collection, storage, distribution, and use of biological samples and associated data. While the importance of public and patient participation has been widely acknowledged in both Western and Chinese academic discussions (e.g., Gaskell and Gottweis 2011; Chen et al. 2013; Lensink et al. 2022), the actual practice of participation differs significantly. Initiatives such as the UK Biobank (Levitt 2005) and the US All of Us Research Program (Kraft et al. 2018) have institutionalized mechanisms, including advisory panels and community engagement, to enhance legitimacy and trust. In China, by contrast, biobank governance remains predominantly state-led and expert-driven. In this top-down context, public participation is designed to align with official regulations, serving to facilitate research rather than provide a bottom-up challenge to state directives. Should a conflict arise between individual preferences and government mandates, particularly regarding national resource security, the latter takes priority, and non-compliant activities are strictly curtailed.

Although studies have examined the Chinese public's perceptions of the ELSI of biobanks (e.g., Chen et al. 2013), the voices of most participants remain largely unheard. They have rarely been meaningfully involved in decision-making, and their potential risks, such as the leakage of sensitive genetic privacy or discrimination by insurance companies based on genetic data, have not been sufficiently addressed. Whether Chinese biobanks can secure sustained societal support will depend on the extent to which participants are meaningfully engaged, not necessarily to challenge existing regulations, but to ensure that biobank governance is perceived as legitimate and trustworthy by the public. Ultimately, the development of biobanks and their regulations is co-constitutive: the rapid growth of biobanking in China necessitated the 2019 Regulations, which in turn provide the legal infrastructure to ensure that biobanks operate in line with national interests.

Beyond public support, factors such as cost-effective biobanking, standardized protocols, sample interoperability, and sustainable funding are critical for the long-term viability of biobanks (Abdaljaleel et al. 2019). China has rapidly expanded its biobanking infrastructure, primarily funded by public agencies such as the Ministry of Science and Technology and universities through short-term (3- to 5-year) projects. Once these projects conclude, however, funding often ceases, leaving many biobanks financially vulnerable. Although policies encourage data sharing, substantial incentives for sharing remain limited, and many collections are used mainly for internal biomedical research. Current Chinese regulations on human genetic resources do not adequately address the commercialization of biobanks, and most institutions have not developed viable business models to manage financial risks. Moreover, many biobanks lack strategic planning and cost analysis; some fail to follow standardized protocols, and some resources are of low quality or lack interoperability. Consequently, sustainability remains a significant challenge for most biobanks in China.

It is widely recognized that transparency, accountability, and oversight are essential for sound biobank governance (Gille et al. 2020). Yet, in China, governance-related information is often inaccessible through publicly available websites and documents. Key questions remain unanswered: How are biobanks operated and overseen? How are samples and data stored, used, and shared? Are internal and external review mechanisms in place? How are research participants' interests protected? What types of informed consent are employed, and through what procedures? What governance objectives, strategies, and mechanisms guide biobank operations? If such questions remain opaque over time, the lack of transparency, accountability, and oversight may erode public trust.

Recent years have witnessed the rapid development of biobanks in China. Although regulators have invited biobankers to participate in expert consultations, and biobankers have contributed to drafting standards, limitations persist within China's biobank governance framework. Governance purposes, objectives, and strategic orientations remain ambiguous, potentially threatening the long-term sustainability of biobanks. Adaptive governance, as discussed earlier, requires continuous learning and resilient systems capable of responding to emerging needs and challenges. In the Chinese context, broader involvement of diverse actors in multi-level decision-making is necessary to develop governance mechanisms that ensure transparency, accountability, and effective oversight.

My fieldwork further reveals a significant gap between the formal regulatory framework, which emphasizes state control, legal codification, and the securitization of genetic resources against perceived external threats, and the practical enactment of these regulations by researchers managing biobanks, which is primarily driven by immediate research needs, collaboration, and sustainability. My long-term engagement in this field, from the early experiences of BIONET to more recent involvement in the BioGovernance Commons, underscores these tensions. Both initiatives provided platforms for cross-cultural dialogue, critical reflection, and mutual learning, creating experimental spaces for governance innovation. For instance, the Commons even invited Jiankui He to participate in a post-incarceration dialogue, transforming a highly sensitive case into an opportunity for collective reflection. Yet such dialogical and experimental approaches remain peripheral within China's broader governance landscape. While currently limited in scale, these bottom-up approaches offer significant value by acting as "regulatory sensors" that can identify ethical tensions and social risks before they escalate into public distrust. Their effectiveness, however, is often constrained by a lack of formal integration into the national decision-making process, meaning they function more as local pilot projects than as drivers of systemic change. The prevailing model continues to rely on top-down regulation and an ethos of national security. For adaptive governance to move beyond a rhetorical aspiration, China must bridge the gap between regulatory formalism and lived practices. Achieving this goal requires not only institutional reforms that promote transparency, accountability, and inclusiveness, but also a cultural shift that recognizes the legitimacy of bottom-up initiatives, patient and public voices, and international collaboration. Only through such dual adaptation of governance mechanisms and regulatory mindsets can China reconcile its self-positioning as a "developing country" with its ambition to achieve global leadership in the life sciences.

References

Abdaljaleel, M., E. J. Singer, and W. H. Yong. 2019. Sustainability in biobanking. *Methods in Molecular Biology* 1897:1–6.

Akyüz, K., G. Chassang, M. Goisauf, et al. 2021. Biobanking and risk assessment: A comprehensive typology of risks for an adaptive risk governance. *Life Sciences, Society and Policy* 17 (1): 10. https://doi.org/10.1186/s40504-021-00117-7.

Argudo-Portal, V., and M. Domènech. 2020. The reconfiguration of biobanks in Europe under the BBMRI-ERIC framework: Towards global sharing nodes? *Life Sciences, Society and Policy* 16 (1): 9. https://doi.org/10.1186/s40504-020-00105-3.

BBCMBA (China Medical Biotechnology Association). 2011. China medical biotechnology association biobank standards (trial). *Chinese Medical Biotechnology* 6 (1): 71–79.

Boeckhout, M., and M. T. Mayrhofer. 2019. Data in question: A survey of European biobank professionals on ethical, legal and societal challenges of biobank research. *PLoS One* 14 (9): e0221496.

Chen, H. 2013. Governing international biobank collaboration: A case study of China Kadoorie biobank. *Science, Technology & Society* 18 (3): 321–338.

Chen, H., H. Gottweis, and J. Starkbaum. 2013. Public perceptions of biobanks in China: A focus group study. *Biopreservation and Biobanking* 11 (5): 267–271.

Chen, Y., C. Sang, Z. Bian, et al. 2021. The scale, collections, and biospecimen distribution of grade a tertiary hospital biobanks in China: A national survey. *Frontiers in Medicine (Lausanne)* 7:560600.

Chu, J., W. Huang, S. Q. Kuang, et al. 1998. Genetic relationship of populations in China. *Proceedings of the National Academy of Sciences* 95 (20): 11763–11768.

Colledge, F., and B. Elger. 2015. Getting a fair share: Attitudes and perceptions of biobank stakeholders concerning the fairness of sample sharing. *Bioethics* 29 (6): 424–430.

Coppola, L., A. Cianflone, A. M. Grimaldi, et al. 2019. Biobanking in health care: Evolution and future directions. *Journal of Translational Medicine* 17 (1): 172. https://doi.org/10.1186/s12967-019-1922-3.

Devriendt, T., M. Shabani, and P. Borry. 2021. Data sharing in biomedical sciences: A systematic review of incentives. *Biopreservation and Biobanking* 19 (3): 219–227.

Gan, R., H. Wang, Y. Song, et al. 2015. Chinese biobanking initiatives. *Biopreservation and Biobanking* 13 (1): 4–7.

Gaskell, G., and H. Gottweis. 2011. Biobanks need publicity. *Nature* 471:159–171.

Gille, F., E. Vayena, and A. Blasimme. 2020. Future-proofing biobanks' governance. *European Journal of Human Genetics* 28 (8): 989–996.

Gottweis, H., and A. Petersen, eds. 2008. *Biobanks: Governance in comparative perspective*. London: Routledge.

Kennedy, A. 2019. China's rise as a science power: Rapid progress, emerging reforms, and the challenge of illiberal innovation. *Asian Survey* 59 (6): 1022–1043.

Kiehntopf, M., and M. Krawczak. 2011. Biobanking and international interoperability: Samples. *Human Genetics* 130 (3): 369–376.

Kraft, S. A., M. K. Cho, K. Gillespie, et al. 2018. Beyond consent: Building trusting relationships with diverse populations in precision medicine research. *American Journal of Bioethics* 18 (4): 3–20.

Lensink, M. A., K. R. Jongsma, S. N. Boers, et al. 2022. Better governance starts with better words: Why responsible human tissue research demands a change of language. *BMC Medical Ethics* 23 (1): 90. https://doi.org/10.1186/s12910-022-00823-7.

Levitt, M. 2005. UK biobank: A model for public engagement? *Life Sciences, Society and Policy* 1:78. https://doi.org/10.1186/1746-5354-1-3-78.

Li, H., M. Ni, P. Wang, et al. 2017. A survey of the current situation of clinical biobanks in China. *Biopreservation and Biobanking* 15:248–252.

Li, Q., L. Jin, and X. Zhou. 2018. A survey of biobank business planning in China. *Medical Journal of Peking Union Medical College Hospital* 9 (3): 271–276.

Meyer, M. 2022. Taking responsibility, making irresponsibility: Controversies in human gene editing. *Social Studies of Science* 52 (1): 127–143.
Ministry of Science and Technology. 2019. Regulations on the administration of human genetic resources. Beijing.
Ministry of Science and Technology. 2011. *Notice on issuing the 12th five-year plan for biotechnology development* (关于印发十二五生物技术发展规划的通知). http://www.gov.cn/zwgk/2011-11/28/content_2005161.htm
Ministry of Science and Technology. 2015. Administrative licensing service guide for the review and approval of the collection, preservation, trade, and export of human genetic resources. Beijing.
Ministry of Science and Technology, et al. 2017a. Notice on issuing the 13th five-year plan for health and medical science and technology innovation (关于印发《"十三五"卫生与健康科技创新专项规划》的通知). https://www.most.gov.cn/tztg/201706/t20170613_133484.html
Ministry of Science and Technology, Ministry of Finance, and National Development and Reform Commission. 2017b. Notice on issuing the plan for optimization and integration of national science and technology innovation bases (科技部 财政部 国家发展改革委关于印发《国家科技创新基地优化整合方案》的通知). https://most.gov.cn/tztg/201708/t20170825_134601.html
Ministry of Science and Technology, and Ministry of Health. 1998. Interim measures for the administration of human genetic resources. Beijing.
Ministry of Science and Technology. 2023. Implementing rules for the regulations on the administration of human genetic resources. Beijing.
Morente, M. M., F. L. Fernandez, E. de Atava, et al. 2008. Biobanking: Old activity or young discipline? *Seminars in Diagnostic Pathology* 25 (4): 317–322.
National People's Congress. 2020a. Biosafety law of the People's Republic of China (amended 2024). Beijing.
National People's Congress. 2020b. Amendment XI to the criminal law of the People's Republic of China. Beijing.
O'Donoghue, S., S. Dee, J. A. Byrne, et al. 2022. How many health research biobanks are there? *Biopreservation and Biobanking* 20 (3): 224–228.
Rijke, J., R. Brown, C. Zevenbergen, et al. 2012. Fit-for-purpose governance: A framework to make adaptive governance operational. *Environmental Science & Policy* 22:73–84.
Rychnovská, D. 2021. Anticipatory governance in biobanking: Security and risk management in digital health. *Science and Engineering Ethics* 27 (3): 30. https://doi.org/10.1007/s11948-021-00305-w.
Shanghai Clinical Research Ethics Committee. 2020. Model for ethics review of human biobanks. *Medicine and Philosophy* 41 (2): 74–80.
State Council. 2012. *Notice on issuing the 12th five-year plan for the development of strategic emerging industries* (国务院关于印发"十二五"国家战略性新兴产业发展规划的通知). Accessed October 3, 2025. http://www.gov.cn/zwgk/2012-07/20/content_2187770.htm.
State Council. 2016. *Notice on issuing the 13th five-year plan for science and technology innovation* (国务院关于印发"十三五"国家科技创新规划的通知). http://www.gov.cn/zhengce/content/2016-08/08/content_5098072.htm.
Sung, W. C. 2009. Within borders: Risks and the development of biobanking in China. In *Human genetic biobanks in Asia: Politics of trust and scientific advancement*, edited by M. Sleeboom-Faulkner, 168–188. New York: Routledge.
Vayena, E., and A. Blasimme. 2021. Towards adaptive governance in big data health research: Implementing regulatory principles. In *The Cambridge handbook of health research*, edited by G. Laurie, E. Dove, A. Ganguli-Mitra, et al., 257–265. Cambridge: Cambridge University Press.
Wang, J. 2018. *The ministry of science and technology publicizes administrative penalties regarding human genetic resources for the first time*; Six institutions including BGI were penalized (科技部首次公开涉及人类遗传资源的行政处罚 华大基因等6家机构遭罚). Beijing News (新京报). Accessed February 26, 2026. https://www.toutiao.com/article/6616220329821864455/
Wang, X., M. Lu, J. Qian, et al. 2009. Rationales, design and recruitment of the Taizhou longitudinal study. *BMC Public Health* 9:223.

Watson, P. H., R. Ravid, C. B. Eng, et al. 2011. What are the main roadblocks to transnational biobank collaboration, and how can we overcome them? *Biopreservation and Biobanking* 9 (3): 213–216.
Zhang, J. 2018. Chinese biobanks: Can't just "save it". *Science and Technology Daily*.
Zhu, S., M. Shen, X. Qiu, et al. 2015. Ethical management guidelines for the Shanghai disease-based biobank network. *Biopreservation and Biobanking* 13 (1): 8–12.

Haidan Chen is a tenured Associate Professor in the Department of Medical Ethics and Law, School of Health Humanities, Peking University (PKU). She is currently an academic visitor at both the Center for Health, Law, and Emerging Technologies (HeLEX) and the Uehiro Oxford Institute, University of Oxford (2026–2027). Prior to joining PKU, she held appointments as a Professor at the College of Humanities and Development Studies, China Agricultural University; a Research Fellow in the Science, Technology and Society Research Cluster at the Asia Research Institute, National University of Singapore (NUS), and a Fellow at Tembusu College, NUS. She was also a visiting postgraduate researcher at the Institute for the Study of Science, Technology and Innovation (ISSTI), University of Edinburgh (2006–2007); a visiting scholar at the Brocher Foundation, Switzerland (June–August 2011); and a Fulbright Visiting Scholar at the Stanford Center for Biomedical Ethics (2016–2017). Chen's research focuses on the ethical, legal, and social implications (ELSI) of genetics and genomics, and the governance of emerging medical technologies in China, particularly stem cells, biobanks, precision medicine, neurotechnologies, and medical artificial intelligence. Her publications have appeared in journals such as *Science*, *Social Science & Medicine*, *BMC Medical Ethics*, and *Neuroethics*.

Chapter 5
Regulation of Cross-Border Transfer of Biological Data in Synthetic Biology Research in China

Zhangyu Wang and Li Du

Abstract The rapid development of synthetic biology (SynBio) has intensified reliance on large-scale biological data, making cross-border transfer of human genetic data (HGD) increasingly central to international research collaboration. In China, however, outbound HGD transfers remain subject to a dense and evolving regulatory landscape shaped by concerns over national security, biosecurity, and data sovereignty. This article examines China's current legal framework governing cross-border HGD transfers and assesses its implications for international SynBio research collaboration. This chapter maps the tripartite regulatory regime governing HGD transfers in China: laws specific to human genetic resources, regulations governing health and medical data, and general data governance laws on cybersecurity and personal information protection. The chapter argues that while China's regulatory emphasis on security and state oversight reflects legitimate historical and geopolitical concerns, the current framework requires recalibration to better support responsible international SynBio research collaboration. Drawing on comparative international practices and the emerging concept of the Biogovernance Commons, the article proposes pathways for harmonizing national security priorities with global scientific cooperation and ethical governance. It concludes that a more coherent and dialogue-driven approach to cross-border HGD governance would enhance China's role in global biodata governance while fostering innovation, trust, and accountability in SynBio research.

Keywords Human genetic data · Personal data protection · Cross-border data transfer · Synthetic biology · International research collaboration · Ethical governance

---

Z. Wang
Faculty of Law, University of Macau, Macau, China

L. Du (✉)
Faculty of Law, University of Macau, Macau, China

Asia-Pacific Academy of Economics and Management, University of Macau, Macau, China
e-mail: stephendu@um.edu.mo

J. Y. Zhang et al. (eds.), *BioGovernance Commons and China's Role in Global Risk Governance*, Emerging Globalities and Civilizational Perspectives,
https://doi.org/10.1007/978-3-032-32198-5_5

Introduction

Synthetic biology (SynBio) is a fast-developing and promising field of biotechnology research and innovation that has been made possible through new discoveries about cellular and molecular processes, advances in DNA computational analyses, and easy synthesis of long strands of DNA. (Cameron et al. 2014; Carlson 2009) With the increasing digitization of DNA and biological data, SynBio research projects can "benefit from burgeoning platform/enabling technologies that would allow large-scale algorithm-driven biological designs, where DNA sequences or parts can be assembled into larger genetic constructs or even genomes or chromosomes" (Bennet et al. 2022; Freemont 2019). On this basis, scientists need to take an immense amount of biological engineering trial-and-error experiments conveniently and cheaply to test large-scale genetic designs (Chen et al. 2020). This trial-and-error experiment is called the "design-build-test-learn" cycle, in which artificial intelligence/machine learning is adopted to record and analyze the relevant biological data, so as to enhance the future design process (Chao et al. 2015).

International research collaborations in the SynBio field are flourishing. The first SynBio project in China, titled "the Programmable Bacteria Catalyzing Research," was funded under the European Commission's Sixth Framework Programme in 2006 (Zhang 2011). In 2019, the Global Biofoundry Alliance (GBA), an alliance covering 32 top-tier non-commercial biofoundries, was formally launched to provide an integrated infrastructure to enable the rapid design, construction, and testing of genetically reprogrammed organisms for biotechnology research and applications (Global Biofoundry Alliance 2023). Along with the increasing international connections, the unobstructed use and free cross-border transmission of biological data are of significance for the research and progress of SynBio.

The cross-border data transfer has never gone without debate in China. Such concerns are rooted in past incidents in which biological samples were transferred by or to foreign research institutes without proper authorization. For instance, in 1997, two epidemiologists from Harvard University took blood samples of over 16,000 farmers in Anhui province for a genetic project without obtaining proper informed consent from the participants (Chu and Liu 2002). After the release of the story, the two scholars were disciplined by the university. This case generated a heated public debate about how to use and share research data when international collaborators were involved. Many Chinese scientists argued that the government should take measures to respond to and counter the impact of foreign biological threats (Chu and Liu 2002; Chen and Song 2018). In October 2018, the Ministry of Science and Technology (MOST) released six administrative penalty decisions on five companies and one hospital, all of which were related to the leakage of human genetic resources (Ouyang et al. 2020; Pu 2018). A 2020 annual report of network security in China revealed that, in 2020, a total of 7.17 million domestic genetic data were found to leave China through the Internet, in which more than 2.73 million were transferred to the US, accounting for 38.1% of the total number of data flow (CNCERT 2021).

Human genetic data (HGD) covers a wide range of information, including data from various levels of biological systems, such as the genome, transcriptome, epigenome, proteome, metabolome, molecular imaging, and molecular pathways (Wörheide et al. 2021). To address potential risks brought by international sharing of HGD, the Chinese government has developed relevant regulations and guidelines. Some examples include the *Administrative Regulation on Human Genetic Resources of the People's Republic of China*, issued in 1998 and amended in 2019 and 2024, and *Guidelines for Administrative Licensing Services for the Collection, Sale, Export, and Exit of Human Genetic Resources*, released in 2015. Scholars have examined China's regulation of HGD cross-border sharing and have found that the sharing of HGD is rigorously controlled by the Chinese government and regulated under a multi-layered framework (Chen and Song 2018; Sharma 2022). Some observers have suggested that such stringent centralized control could pose challenges for international research cooperation (Mallapaty 2022; Sharma 2022). These strict rules may discourage some potential collaborative research, introduce additional uncertainties for ongoing projects with regard to data sharing, and potentially lead to a degree of isolation from the international research community (Mallapaty 2022; Sharma 2022).

There has been a growing consensus that the free flow of HGD would benefit the advancement of genomics-based biomedical research (Knoppers 2014). A balanced approach that facilitates cross-border HGD sharing, rather than relying solely on legal restrictions, may better serve the goal of fostering international research collaborations. In this context, the concept of "Biogovernance Commons" offers a promising transnational approach. Launched during the Covid-19 pandemic by several leading bioethics scholars, including Joy Y. Zhang (UK) and Ruipeng Lei (China), this initiative emphasizes collaborative and common practices, i.e., building shared understanding and cooperative governance mechanisms for life sciences between China and the West (Zhang et al. 2022, 2023). The project advocates for developing more inclusive bottom-up dialogue that serves as complementary channels to facilitate China's regulatory system on issues such as bioethics, data sovereignty, and equitable research collaboration, particularly in fast-evolving fields like synthetic biology (Zhang et al. 2022, 2023). Accordingly, in this study, we undertake a close examination of China's current regulatory framework for cross-border HGD transfers, aiming to examine whether the current legal framework on HGD cross-border transfers functions well to cater for international SynBio collaborative research. Our analysis starts off by exploring laws applicable to HGD cross-border transfers. We then examine conflicts and regulations that require revisions in the current regulatory framework. We argue that China should play an active role in developing an international governance model on HGD sharing that could help harmonize national security priorities with global scientific cooperation and ethical standards. To foster research advancements in SynBio, China should promote international governance dialogues on HGD sharing and establish a more open and efficient management system for cross-border HGD transfers.

Laws Governing Cross-Border Transfer of HGD

China currently has a tripartite legal framework for governing cross-border transfer of HGD, including laws governing biological research, laws regulating the health and medical data, and laws applicable to the general data governance (CNCERT 2021). The oversight is further complicated by the lack of direct guidance by the government as to which laws should be prioritized in regulating the cross-border flow of the HGD for research purposes.

Laws Governing Human Genetic Research Data

The Chinese government has identified HGD as a national strategic resource and emphasized the importance of good governance practices relevant to it (Chen and Song 2018). In this regard, the *Administrative Regulation on Human Genetic Resources of the People's Republic of China* (*HGR Regulation*) was amended in 2019, aiming to systematically strengthen the standardized management of Chinese HGD and encourage its sound use. In the SynBio research domain, cross-border flow of HGD may occur in two scenarios: (1) Chinese entities providing or making Chinese HGD available to foreign entities (direct cross-border transfer) and (2) cross-border transfer of HGD in an international collaborative project (Art. 21 and 28, the HGR Regulation).

1. Direct Cross-Border Transfer of HGD

   The current regulations prohibit foreign entities from collecting or preserving HGD within China or providing HGD to other entities abroad. According to the *HGR Regulation*, filling for the record (in Chinese: 备案) is the prerequisite for the HGD to be legally transferred overseas. Specifically, those who provide or offer access to HGD to foreign entities shall file for the record to the National Health Commission of the People's Republic of China (NHC) and submit such data for a backup (Art. 28, the HGR Regulation). Additionally, if the transmission of such HGD poses threats to public health, national security, and the public interest of society, an NHC-organized security review may be initiated (Art. 28, the HGR Regulation). However, the newly established *Biosecurity Law*, which entered into force in 2021, sets up different requirements for providing Chinese HGD to overseas organizations or individuals. According to the *Biosecurity Law*, those who provide or make Chinese HGD available to foreign entities shall "report in advance" (in Chinese: 事先报告) and "submit a backup of the HGD" to the NHC (Art. 57, the Biosecurity Law). Moreover, the *Biosecurity Law* does not mention a security review process for a cross-border HGD transfer application that involves potential risks to public health, national security, and the public interest.

2. Cross-Border Transfers of HGD in International Collaborative Projects

In an international collaborative project, the cross-border transfer of HGD between Chinese and foreign research teams is inevitable. To enable such collaboration, Chinese and foreign cooperators should sign a written agreement to illustrate related cooperative matters and jointly apply for approval from the NHC on such collaborative project (Art. 22, the HGR Regulation). According to the *HGR Regulation*, when deciding whether to approve an international collaborative project, the NHC should review whether the HGD used therein is legally acquired, and whether the types and quantities of such HGD are consistent with the research contents (Art. 22, the HGR Regulation).

Laws Governing HGD Generated in the Provision of Healthcare Services

The management of HGD may fall within the scope of the NHC's supervision authority if the HGD is generated from healthcare and medical treatment. The *Measures on Population Health Information* (*PHI Measures*) is currently the main applicable regulation on the collection and processing of population health information acquired in the healthcare sector. According to the *PHI Measures*, the population health information covers basic population information and medical and health service information, and HGD should be recognized as a type of population health information (Art. 3, the PHI Measures). Accordingly, health and medical institutions need to comply with the *PHI Measures* to demonstrate their legal compliance in HGD dispositions. However, while encouraging and promoting the nationwide connectivity, timely update, and sharing of population health information, the *PHI Measures* prohibit healthcare institutes from storing the population health information in overseas servers (Art. 10 and 13, the PHI Measures). This data localization requirement may limit the mobility of HGD generated from healthcare institutes in China. In addition, the *Measures for the Management of Health Big Data* (*Healthcare Big Data Measures*) released in 2018 define healthcare-related big data as the data "generated in the course of disease prevention and control, as well as health management" (Art. 4). It requires that healthcare-related big data, in principle, should be stored in safe and reliable servers in China. If healthcare-related big data needs to be transferred abroad, according to the *Healthcare Big Data Measures*, the application should undergo a security assessment and review in accordance with relevant laws and regulations (Art. 30).

Laws Applicable to General Data Governance

Cross-border HGD transfers are also governed by laws related to general personal data protection and cybersecurity protection. In recent years, the Chinese government has enacted laws and administrative regulations to strengthen the governance of cybersecurity and to address increasing concerns about personal data protection.

For example, the *Cybersecurity Law of the People's Republic of China* (The *Cybersecurity Law*) and the *Data Security Law of the People's Republic of China* (The *Data Security Law*) were enacted to regulate data processing activities and improve cyber and data security. The *Personal Information Protection Law of the People's Republic of China* (The *PIPL*) focuses on protecting the right and interest of personal information (Art. 1, the PIPL). Under these laws, HGD can be categorized into three data types, namely, personal information, sensitive personal information, and critical data. For example, according to the *PIPL*, personal information encompasses all information related to identified or identifiable natural persons recorded by electronic or other means, excluding information processed anonymously (Art. 4, the PIPL). Accordingly, HGD deriving from human cells, unless anonymized, falls within the scope of the definition of personal information in the *PIPL*. Moreover, HGD that reveals biometric identification or medical health of certain people is considered sensitive personal information according to the *PIPL* and hence receives a higher-level protection (Art. 28, the PIPL). Furthermore, HGD can also be classified as critical data under the *Cybersecurity Law* if its illegal disclosure or abuse is likely to pose threats to national security, economic performance, social stability, and public health and safety (Art. 19, the Security Assessment Measures for Outbound Data Transfers).

The *PIPL* has adopted a "triple safeguards" model to regulate cross-border data transfers. Based on Article 38 of the *PIPL*, if an HGD data processor intends to provide personal information to a party outside China, it can choose to (1) apply for and pass security assessment by the national Cyberspace Administration, (2) be granted personal information protection certification by a specialized institution, or (3) enter into a contractual relationship with the foreign party by signing the standard contract agreements developed by the national Cyberspace Administration. Under Article 38 of the *PIPL*, these three models in principle run parallel, and data processors need only opt for one to accomplish legal compliance when transferring data overseas. In some cases, when the volume of data transferred outbound by a data processor exceeds a specified threshold, a security assessment is mandatory for the data cross-border transfer, and, in this case, the other two options are not allowed to be used (Art. 40, the PIPL). Before proceeding with a security assessment application to the national Cyberspace Administration, the HGD processor should proceed with a self-assessment and then prepare a declaration form, self-assessment report,[1] and legal documents (contracts or others) signed by the data receiving body

[1] According to Article 5 of the *Security Assessment Measures for Outbound Data Transfers*, matters need to be examined in the self-assessment include: (1) the legality, legitimacy and necessity of the purpose, scope and method of the outbound data transfer and data processing by the overseas recipient; (2) the scale, scope, type, and sensitivity of the data to be provided abroad, and the risks to national security, public interests, or the legitimate rights and interests of individuals or organizations caused by the outbound data transfer; (3) the responsibilities and obligations that the overseas recipient promises to undertake, and whether the overseas recipients' management and technical measures and capabilities for performing its responsibilities and obligations can guarantee the security of the outbound data transfer; (4) risks of the data to be tampered with, destroyed, divulged, lost, transferred, illegally obtained, or illegally used during and after the outbound data

for submission (Art. 6, the Security Assessment Measures for Outbound Data Transfers). These materials will later serve as the basis and evidence for the national Cyberspace Administration's review and assessment. In March 2024, the *Provisions on Promoting and Regulating Cross-border Data Flows* issued by the Cyberspace Administration introduced conditions and distinctions for the selection among the three models (Arts. 7–8).[2] Under the 2024 *Provisions*, when the triggering conditions for mandatory declaration for a security assessment are not satisfied, HGD processors can only undergo either of the other two alternative paths. Consequently, the three paths are no longer parallel and optional.

Legal Challenges of Regulating HGD Cross-Border Transfer

As illustrated above, China currently adopts strict and complex regulations on the cross-border transfer of HGD. These laws do not allow foreign entities to directly transfer HGD from China to other countries (Art. 7, the HGR Regulation). Therefore, foreign entities ought to seek the assistance of or cooperation with Chinese partners for the intended cross-border HGD transfer. Chinese parties, who assist or collaborate with the foreign entities, according to laws and regulations, should substantially participate in the whole research process, which is subject to the rigid government's approval and supervision (Art. 21, the HGR Regulation). Previous studies indicate that rationales for such requirements are likely based on considerations that (1) through substantial participation, Chinese researchers may obtain scientific experience, skills, and funding support from foreign collaborators, and (2) substantial participation will safeguard Chinese national interests and security by avoiding foreign countries' exploitation of Chinese human genetic resources. (Mu and Yu 2021;

transfer; whether the channel for maintenance of personal information rights and interests is smooth; (5) whether the relevant contracts on the data to be concluded with the overseas recipient or other legally binding documents have fully agreed on the responsibilities and obligations to protect the data security; and (6) other matters that may affect the security of the outbound data transfer.

[2] According to Article 7 and 8 of *Provisions on Promoting and Regulating Cross-border Data Flows*, data processors fitting into any following category should apply for the security assessment organized by the Cyberspace Administration of China: (1) Where a critical information infrastructure operator provides personal information or important data abroad; or (2) Where any data processor other than a critical information infrastructure operator provides important data abroad or, as of January 1 of the current year, provides personal information (excluding sensitive personal information) of not less than 1 million people or sensitive personal information of not less than 10,000 people in aggregate to overseas parties. Where any data processor other than a critical information infrastructure operator provides abroad the personal information (excluding sensitive personal information) of not less than 100,000 but not more than 1 million persons, or the sensitive personal information of not more than 10,000 persons, accumulatively as of January 1 of the current year, it **shall** conclude a standard contract with overseas recipients for provision of personal information abroad or go through the authentication on protection of personal information in accordance with the law.

Tou and Huang 2015). While the current regulatory framework features robust state controls over HGD management, it also contains certain ambiguities and uncertainties. Addressing these gaps could help reduce operational difficulties and legal hurdles in practice, thereby facilitating more effective cross-border HGD transfers. We identified three primary challenges that may negatively impact the smooth HGD cross-border transfers and the advancement of cross-border SynBio research collaboration.

Legislative Incoherence and Overlapping Requirements

Different Requirements Between the *Biosecurity Law* and the *HGR Regulation*

The inconsistent requirements set in the *Biosecurity Law* and *HGR Regulation* for the cross-border flow of HGD have resulted in much confusion in practice. The *HGR Regulation* requires that an HGD processor file for record and submit pertinent data backup for the intended HGD cross-border transfer, while this "filing for record" clause was later presented as a "report in advance" requirement under the *Biosecurity Law*. The duty to "report in advance" highlights the timing of fulfilling such a procedural obligation, but it is not clear whether this procedure would give NHC an extended authority to evaluate the report and make a decision regarding the approval of proposed cross-border transfers. Moreover, although Article 20 of the *Biosecurity Law* states that the government shall establish a biosecurity review regime with aims to effectively preventing and eliminating biosecurity risks, it has not clearly outlined the process for a security review of cross-border HGD transfers as required under Article 28 of the *HGR Regulation*.

In May 2023, the "*Implementing Rules for the Administrative Regulation on Human Genetic Resource* (The *Implementing Rules for the HGR Regulation*)" came into effect, specifying the procedural arrangement for the proposed HGD cross-border flow. According to the *Implementing Rules for the HGR Regulation*, any domestic entity providing, or opening access to, HGD to a foreign entity shall "report in advance" to the NHC, and the materials submitted for report shall contain the statement of HGD sharing purpose, the specific content of HGD, information of the HGD receiving entities, and the assessment of potential risks posed to domestic protection of human genetic resources in China (Art. 36). The security review will be initiated if the intended HGD flow endangers China's public health, national security and societal interests. In addition, the *Implementing Rules for the HGR Regulation* have identified the types of HGD of which the cross-border sharing should be subject to compulsory reviews: (1) HGD of important genetic families, (2) HGD acquired in specific areas, and (3) exome sequencing and genome sequencing information resources of a population of more than 500 persons (special HGD) (Arts. 37–38). While the *Implementing Rules for the HGR Regulation* postulates practical rules for implementing the *HGR Regulation* and clarifies some ambiguities, it does not address legislative contradiction issues related to the requirement of "report in advance" laid down in the *Biosecurity Law* and the "filing for record" in the *HGR Regulation*.

Incoherence Between the PHI Measures and the HGR Regulation

Rules of the *PHI Measures* have raised legal hurdles for realizing a free HGD flow, as it requires that all population health information obtained by medical organizations should be stored in domestic servers (Art. 10). In particular, the *PHI Measures* is widely applied to medical and health organizations at all levels and regions in China and, even, to those institutions providing traditional Chinese medical treatments (Art. 2). Such data localization requirement consequently increases the difficulties for implementing international HGD transfers and contradicts the spirit of the laws outlined in the *HGR Regulation*, i.e., that the Chinese government should encourage medical institutions to carry out international cooperative scientific research by making use of Chinese HGD, so as to enhance national competitiveness in relevant research and development capabilities (Art. 19, the HGR Regulation). The incoherence between the two regulations is likely to increase the legal compliance burdens for SynBio research that involves Chinese HGD and relates to medical applications.

Various and Unclearly Regulated Assessment Requirements

1. Assessment Requirements under the PIPL

As explored above, the *PIPL* has established a mode for regulating personal information cross-border flow, i.e., passing the security assessment, concluding a standard contract, and receiving a certification. Governments also developed administrative rules to guide the implementation of the laws. For example, the *Security Assessment Measures for Outbound Data Transfers* have been established to guide how to implement security assessments, and the *Measures for the Standard Contract for Outbound Transfer of Personal Information* (The *Standard Contract Measures*) came into effect in June 2023. The latter includes provisions about when a standard contract should be used and specifies the content a standard contract should include. However, there is an overlap in requirements between these two rules. Specifically, according to the *Standard Contract Measures*, HGD processors opting for concluding a standard contract to transfer HGD should submit an assessment report regarding the data transfer's impacts on personal information protection to the national Cyberspace Administration (Art. 7, the Standard Contract Measures). Article 5 of the *Standard Contract Measures* states what should be assessed and what content should be regulated. The provisions are similar to those in Article 8 of the *Security Assessment Measures for Outbound Data Transfers* (See Table 5.1). Given that the purpose of using a standard contract is to simplify and facilitate the process of data cross-border transfer, the requirement for submitting an assessment report of the data transfer's impacts on personal information protection might not alleviate the compliance burden on HGD processors as legislators envision (Zhao 2023). In our opinion, the current *Standard Contract Measures* are unlikely to be effective in streamlining relevant legal procedures and promoting data cross-border flow.

2. **Between the** *HGR Regulation* **and the** *PIPL*

When an HGD cross-border transfer application is deemed to pose threats to the public health, national security, and public interest, both a security review (安全审查), required by the *HGR Regulation*, and a security assessment (安全评估), requested by the *PIPL*, shall be conducted. However, it is unclear which rules should be prioritized or whether a collective review should be carried out. For the security review required by the *HGR Regulation*, the provision only states that an HGD processor should submit the materials to the NHC for a security review, but no specific rules are established about how to carry out the security review (Art. 28, the HGR Regulation). *The Implementing Rules for the HGR Regulation* have also not developed practical guidance regarding the security review, but rather only stated briefly that the NHC should organize experts to conduct the corresponding review (Art. 40). In addition, regardless of circumstantial self-assessment requirements being set in the *PIPL*, provisions concerning self-assessment alike have not been seen in the *HGR Regulation*. Therefore, it is not clear whether the self-assessment is needed as a part of the security review under the *HGR Regulation*. The overlapping security assessment requirements with large uncertainty in rulemaking can subject the processors to onerous and multiple legal procedures for demonstrating legal compliance, consequently delaying the progress in SynBio research.

Lack of Clarity in Guiding International Collaborative Projects

The NHC has the authority to review the acquisition of HGD and research content for international cooperation projects. However, it is not clear whether the cross-border transfer of HGD in an international collaborative project needs to fulfil procedural requirements, that is, filing for record and submitting a backup. As the former authority in charge of HGD administrative management, the MOST's answers to this question on its official website are not sufficient to dispel the confusion. In the *FAQ Regarding the Management of Human Genetic Resources* (The *First FAQ*) published on 04 March 2022, the MOST stated that providing HGD to organizations other than the participating parties requires the procedure of filing for

Table 5.1 Comparison between impact assessment on personal information under standard contract measures and security assessment under the security assessment measures for outbound data transfers

	Measures for the Standard Contract for Outbound Transfer of Personal Information (Standard Contract Measures)	Security Assessment Measures for Outbound Data Transfers
G34 Specific contents of legal provisions	Article 5: Prior to the outbound transfer of personal information, the personal information processor shall conduct a personal information protection impact assessment, with the focus on the following: (1) the legality, propriety, and necessity of the purposes, scope, and methods of the processing of personal information by the personal information processors and foreign recipients; (2) the volume, scope, type, and degree of sensitivity of the personal information transferred and the potential risks to the rights and interests in personal information that might be brought; (3) the obligations that the foreign recipient has pledged to bear, as well as whether the management and technical measures and capacity for the performance of obligations can ensure the security of the transferred personal information; (4) the risk of personal information being altered, damaged, leaked, lost, and abused, and so forth after it is transferred, and whether the channels for individuals to exercise their personal information rights and interests are accessible and smooth; (5) the impact of the policies, laws, and regulations of the foreign recipient's nation or region on the performance of a standard contract; (6) other matters that might impact the security of transferred personal information.	Article 8: Data transfer security assessments are to focus on assessing the potential risks to national security, public interests, and the lawful rights and interests of individuals and organizations that might be brought on by the transfer of data, primarily including the following matters: (1) the legality, propriety, and necessity of the purpose, scope, and methods, etc., of transferring the data; (2) the impact of the data security protection policies and regulations and the cybersecurity environment of the country or region where the overseas recipient is located on the security of data to be transferred abroad, and whether the data protection level of the overseas recipient meets the requirements of the laws and administrative regulations of the People's republic of China and mandatory national standards; (3) the scale, scope, types, and sensitivity of the transferred data, and risks during and after transfer, such as the data being altered, destroyed, leaked, lost, or transferred, or being illegally obtained or used; (4) whether data security and rights and interests in personal information can be fully and effectively safeguarded; (5) whether the legal documents that data processors will conclude with the overseas recipients fully provide for the responsibilities and obligations in protecting data security; (6) compliance with Chinese law, administrative regulations, and departmental rules; (7) other matters that the National Cyberspace Administration finds necessary to be assessed.

record and backup (MOST 2022a). However, a month later, the MOST added the *FAQ Series 2*, stating that HGD produced by the collaborative research using Chinese HGD can be used by both parties without filing for the record. If other participating foreign entities want to use the relevant research data, the Chinese party shall file for record, but the *FAQ Series 2* has not defined what constitutes the "research data" (MOST 2022b). Importantly, these two *FAQs* have not explained whether other foreign units participating in international collaborative scientific research that want to use HGD need to file for a record. In 2025, the NHC released a series of *FAQs regarding the Management of Human Genetic Resources*; however, this issue remains unaddressed in these FAQs (NHC 2025). This uncertainty further raises concerns about whether such cross-border transfers of HGD to a foreign unit participating in the international collaborative scientific research would trigger a security review, since, according to the *HGR Regulation,* the NHC should premise its approval on international collaborative projects upon the condition that there is no harm to the public health, state security, or public interest of China (Art. 22).

Informed Consent Procedures

The *HGR Regulation* and the *PIPL* generally require data processors to inform HGD providers and obtain their consent to transfer their HGD outside China. The *HGR Regulation* stipulates that informed consent is the precondition to accomplish an HGD cross-border transfer, with adherence to the protection of privacy and legitimate rights of HGD providers (Art. 9). Except for this general requirement, there are no detailed provisions in the *HGR Regulation* to guide how informed consent should be obtained. It is worth noting that although the *HGR Regulation* does not specify the informed consent procedure, it refers the supervision of the informed consent procedure to the ethical committee for review and approval (Art. 9). The *Measures for the Ethical Review of Biomedical Research Involving Humans* (The *Ethical Review Regulation*) regulate ethics review in China, where informed consent is recognized as a fundamental ethics principle and human right (Art. 18) (Yu and Li 2014). Requirements in the *Ethical Review Regulation* urge HGD processors to convey the information in a complete and understandable manner and then obtain the consent legally and appropriately. More specifically, the *Ethical Review Regulation* has proposed informed consent matters related to HGD use in biomedical research, such as the use of appropriate language, elements of consent and consent re-acquisition, special protection to vulnerable groups, rights of HGD providers, and informed consent exemptions (Arts. 33–39). However, the *Ethical Review Regulation* does not clarify what informed consent procedures should be performed in HGD cross-border transfer scenarios, but only stipulated that the ethics review in international research collaboration should be completed by the ethics committees of the participating Chinese organizations (Art. 30). Furthermore, the scope of the *Ethical Review Regulation* is rather limited as the measures are not applicable to biomedical research that takes place outside of medical settings (Art. 2). To address this gap, amended *Ethical Review Regulation* was drafted in 2021 and issued for public comments (NHC 2021). Article 2 of this proposed amendment requires that all life science and medical research activities involving human beings should be subject to ethical review.

According to the *PIPL*, in order to transfer HGD overseas, an HGD processor should obtain separate consent from the HGD provider after informing them of matters legally required (Art. 39). However, the application of "separate consent" has caused uncertainties in practice. More specifically, the *PIPL* has not clarified whether the attribute of "separate consent" differentiates from "consent", and if that is the case, whether "separate consent" equates with "additional consent" or "re-consent." It further raises the question that, for the HGD having been obtained by legitimate means (e.g., by securing an appropriate level of informed consent and in accordance with statutory requirements and the public interest), whether the HGD processor needs to obtain consent again for further HGD processing (cross-border transfer) (Yang 2022). In addition, the *PIPL* only outlines the general terms of the informed consent procedure, without specifying how it will be implemented, thereby raising questions about its effectiveness in comprehensively informing HGD providers.

Recommendations for Improvements and Conclusion

In our previous discussion, we presented the landscape of current regulation of HGD cross-border transfers and revealed the legal challenges that may hinder the free flow of HGD. In this section, we will conclude this study by proposing some recommendations for the challenges identified.

Addressing the Complexities of the Current Regulatory Framework

Our analysis indicates that the current legal rules related to the administration of HGD in China are multi-layered and complex, and in some instances present certain inconsistencies. Several administrative agencies are responsible for assessing and overseeing HGD transfer activities. In practice, this multifaceted regulatory framework can impose significant compliance burdens and create challenges for research teams working on projects that involve large volumes of personal information, including HGD (Mallapaty 2022; Lewis 2023). We suggest that legislature or administrative agencies should be more proactive in addressing legal gaps and incoherence in the governance of HGD, while at the same time continuing to promote the interoperability of various rules contributing to the efficiency and effectiveness of HGD cross-border flow. To achieve this goal, the legislature and government authorities should actively engage with stakeholders such as HGD processors, research teams, and legal experts to gain a better understanding of the current law's implementation and actual societal needs and inform future legislation improvement.

Improving the Informed Consent Process

The informed consent procedure for HGD cross-border transfer should be improved by developing detailed and timely updated rules and guidelines that better facilitate HGD processors' implementation of informed consent processes and protection of participants' privacy. Traditionally, specific informed consent is widely used, which requires HGD providers to be informed that their HGD will be transferred overseas for specific experiments or research, and to give consent on this basis (Helgesson 2012). However, it is possible that HGD flow or future research will not follow the intended interventions and purposes initially outlined in consent documents signed by HGD providers. In practice, many countries, in preserving the specific informed consent, are constantly moving toward "broad consent" to biomedical data in research use. For example, the EU in the Recital 33 of the *General Data Protection Regulation* (The *GDPR*) has recognized the limitations of obtaining specific consent in the data collection stage for subsequent research and has stipulated that the data subject has the right to consent to "certain areas of scientific research." Australia also provides the possibility of obtaining "board consent" and "blank consent" to the use of data in future research, which in the *National Statement on Ethical Conduct in Human Research* (Arts. 2.2.14–2.2.18) are termed as "extended consent" and "unspecified consent," respectively. Although this *National Statement* is not legally binding, it is still widely used and highly respected since researchers are only eligible to apply public funding upon declaring compliance with it (Eckstein et al. 2018). In a public consultation for future reforms in data protection laws, the UK is considering further clarifying the use of broad consent that has already existed in the UK data protection law, in order to "reduce uncertainty and concerns around misuse" (DCMS 2022).

The Chinese *Ethical Review Regulation* stipulates that if biological sample donors consent in writing that their donated samples will be used in all medical research, researchers can utilize their HGD without reconsenting, upon approvals by ethical committees (Art. 39). However, this alternative approach to securing broader level of consent, as many Chinese scholars argued, is ambiguous and unlikely to truly benefit biological research (Li et al. 2022; Zhang et al. 2017; Chen et al. 2018). Overall, China is still firmly on the path of specific consent. Therefore, we recommend that specific consent should primarily be preserved as the "standard requirement" to target those HGD providers that need more clarity and predictability for their HGD future use (Helgesson 2012). In addition, China should consider introducing new informed consent strategies, adapting to the vast flow and fast circulation of HGD currently brought by the advanced networks and digital technologies. For example, broad consent offering convenience to researchers should be formally launched and continuously improved in legislation to a well-functioning state, along with strengthening public education to accept such an approach in HGD provision.

Dynamic consent is also an option worthy of considering, which refers to a personalized consent process through ongoing dialogues between researchers (HGD processors) and research participants (HGD providers) on network platforms (Kaye et al. 2012). In a dynamic consent scenario, HGD providers can communicate with HGD processors in real time and can choose whether to consent to the new use in a personalized way (Budin-Ljøsne et al. 2017). Different from broad consent, dynamic consent provides HGD providers with greater choice flexibility, whereas broad consent is intended for the new and unforeseeable HGD use in future research. By using a broad consent approach, participants will only be asked to re-consent for those "non-trivial" reasons that have a substantial deviation from the consent originally given (Prictor et al. 2020; Kaye et al. 2015; Steinsbekk et al. 2013).

Various consent strategies have corresponding application scenarios, so China should proactively determine alternatives in facilitating HGD cross-border flow and diverse research needs. Meanwhile, institutional and technical restrictions and safeguards should be developed, contingent on a specific informed consent strategy, in order to ensure that researchers do not abuse these alternative consent strategies with more flexibility for their own research convenience.

Further Considerations Regarding the Regulation of Trans-Border HGD Transfer

In addition to clarifying relevant rules on security assessment and informed consent procedures, it is also important to consider what kind of regulatory regime China should establish that can facilitate safe HGD transfers for research purposes in the context of SynBio research. Our analysis of the existing legal framework indicates that China has established robust control over the cross-border flow of HGD with the goal of protecting national biological and data security (Wu 2022; Ma and Liu 2022). China has adopted the regulatory model of using "localized data storage as the principle with cross-border flow as an exceptional case under security assessment" (Zhang and Tu 2021; Huang and Wei 2021). Under the current highly regulated environment, researchers do not enjoy any privileges in transferring HGD overseas for research use. In particular, the existing multi-layered legal framework presents researchers, as well as other HGD processors, with various legal barriers and heavy burdens to achieve legal compliance. This inevitably leads to inefficiency and unnecessary obstructions to international HGD transfer.

China's 2021 "14th Five-Year" Plan on Bioeconomy Development underscores the importance of participating in global biosecurity governance and promoting bilateral and multilateral international cooperation in the life sciences and biotechnology (NDRC 2021). While this policy orientation reflects a strong top-down commitment to international engagement, the effectiveness of such an approach ultimately depends on how it is translated into a collaborative and forward-looking governance mechanism that incorporates broader societal and professional input. One possible avenue for China to substantiate such a mechanism would be to more actively participate in transnational governance dialogues on HGD sharing, particularly by encouraging bottom-up engagement from scientific communities, professional associations, and public stakeholders. As the "Biogovernance Commons" initiative advocates, scientists, ethicists, policymakers, and public stakeholders across borders should be engaged in inclusive and iterative dialogue to foster shared understanding and help develop globally relevant ethical and governance standards for emerging biotechnologies (Zhang et al. 2022, 2023). Such engagement would enhance mutual trust in international research collaborations, promote adaptive governance models that keep pace with technological advances, and strengthen China's role as a responsible leader in global biodata governance.

China could establish a national advisory body or a cross-ministerial working group dedicated to HGD governance, ensuring that domestic policies on transnational HGD sharing are both informed by and contribute to global discussions on synthetic biology ethics, data justice, and equitable innovation. Such a body could serve as a coordination and support platform, which interconnects government regulators with scientists, ethicists, legal scholars, and broader stakeholders. Sufficient institutional support for such bottom-up engagement would further strengthen mutual trust in international research collaborations and position China as an active and responsible participant in biogovernance.

China can also draw lessons from the international community's efforts to develop an open and efficient regime for the cross-border flow of HGD. For example, the OECD released the *Guidelines for Human Biobanks and Genetic Research Databases* in 2009, recommending that the operators of the bio-databases should strive to make data and materials rapidly and widely available to researchers to advance knowledge and understanding (Art. 1.C). In 2013, the *OECD Guidelines on the Protection of Privacy and Transborder Flows of Personal Data* urged its member states to adopt reasonable and appropriate measures to ensure fluent and secure transborder flows of personal data, and member states should not obstruct such flows in the name of the protection of privacy and individual liberties. (Part Three, Art. 16 and 18). In 2017, the OECD released its *Recommendation on Health Data Governance*, particularly reiterating the need for member states to take active steps to support international cooperation in processing personal health data for health system management, research, statistics, and other health-related purposes (Art. IV). Many international projects committed to genetic research, such as ICGC (International Cancer Genome Consortium), GA4GH (Global Alliance for Genomics and Health), have also emphasized their visions of facilitating international free flow and responsible sharing of HGD through their continuous efforts in order to advance scientific research, public health, and prosperity.[3]

[3] *See*, for example, the vision of International Cancer Genome Consortium (ICGC): the ICGC intended to formulate a comprehensive cancer genomics atlas covering all major tumor types and subtypes that are of main importance across the globe, based on which it makes global genomic

In practice, the UK Biobank offers a leading model for promoting open science in the public interest. Launched in 2012, it operates as an open-access resource, removing restrictive preconditions of research collaborations for foreign scientists to use domestic HGD (Sudlow et al. 2015). Since 2012, the UK Biobank has been streamlining registration and approval procedures for researchers to acquire HGD used both in domestic and international research projects (Conroy et al. 2019). Benefits from such open international HGD sharing are substantial, and free access to genetic data offered by the UK Biobank has attracted researchers around the world and has greatly promoted research and innovation in both academia and industry (Conroy et al. 2019).

Moreover, new guidelines for accelerating the processing of personal data for research purposes have been developed to maintain and further the UK's international leadership in genetics research (Zhang 2022; Szustakowski et al. 2021). For example, the *Genome UK: The Future of Health Care* released in 2020 established the policy foundation for genetic research in UK for the next decade, in which it emphasized the country's commitment to "working internationally to advance the shared understanding of genomics" and to developing feasible "protocols and standards that ensure research-friendly data accessing and re-using" (Department of Health and Social Care et al. 2020). In 2021, the Department for Digital, Culture, Media & Sport published a document for public consultation on reforms to the UK's data protection regime—*Data: A new direction*, in which a special regime was proposed to reduce legal barriers for responsible scientific research using personal data (DCMS 2021). Later, in the *Data: A new direction—government response to consultation*, the Department articulated the importance of data free flow in supporting the development of "cutting-edge data-driven technologies" (DCMS 2022). In the first chapter of the *government response*, six aspects regarding the directions of future reforms on the data protection regime were illustrated, so as to foster "clear and consistent rules for the use of personal data" in scientific research (DCMS 2022). These six aspects included bringing further clarity to "personal data used for research purposes" and "data further processing (re-using)," making arrangements related to interaction between AI technologies and the UK's data protection regime, and proposing the "data intermediaries" as a way to responsibly share the confidential personal data, among others (DCMS 2022).

Following these international examples, China could consider continuing to align with the international trend toward research data sharing and take on a leadership role in developing paradigms for open and responsible sharing of scientific data

data sharing for cancer possible by providing the international research community with comprehensive genomic data for many cancer types (Available at https://www.icgc-argo.org/page/64/about-icgc-argo Accessed 15 March 2026.); The Global Alliance for Genomics and Health (GA4GH) (2019) Framework for responsible sharing of genomic and health-related data: The GA4GH established in 2013 aims to develop a data sharing framework that possesses "harmonized approaches to enable effective and responsible sharing of genomic and clinical data and to catalyze data sharing projects that drive and demonstrate the value of data sharing." (Global Alliance for Genomics and Health. Available at https://www.ga4gh.org/framework/ Accessed 15 March 2026).

(Peng et al. 2016). China should promote dialogues on HGD sharing governance at the international level, and the legislative reforms and biobank management approaches adopted in the UK offer valuable lessons for enhancing China's own regulatory framework in synthetic biology and broader scientific research. Specifically, China could balance openness with security by advancing encryption and other technical safeguards, appropriately relaxing controls over cross-border HGD flows for research purposes, while simultaneously refining informed consent guidelines and simplifying administrative and security assessment procedures. Through such measures, China could foster an ecosystem that supports both innovation and ethical governance in the global scientific community.

Acknowledgement This research is funded by the University of Macau, the Asia-Pacific Academy of Economics and Management under the Seed Research Grant No. APAEM/SG/0002/2025.

References

Bennet, Devasier, Tuan Vo-Dinh, and Frederic Zenhausern. 2022. Current and emerging opportunities in biological medium-based computing and digital data storage. *Nano Select* 3:883–902.

Budin-Ljøsne, Isabelle, J. A. Harriet, Jane Kaye Teare, Stephan Beck, Heidi Beate Bentzen, Luciana Caenazzo, Clive Collett, Flavio D'Abramo, Heike Felzmann, and Teresa Finlay. 2017. Dynamic consent: A potential solution to some of the challenges of modern biomedical research. *BMC Medical Ethics* 18 (1): 1–10.

Cameron, D. Ewen, Caleb J. Bashor, and James J. Collins. 2014. A brief history of synthetic biology. *Nature Reviews Microbiology* 12 (5): 381–390.

Carlson, Robert. 2009. The changing economics of DNA synthesis. *Nature Biotechnology* 27 (12): 1091–1094.

Chao, Ran, Yongbo Yuan, and Huimin Zhao. 2015. Building biological foundries for next-generation synthetic biology. *Science China Life Sciences* 58 (7): 658–665.

Chen, Xiaoyun, Yu Tian, Li Ping, Liang Dong, Liang Shen, and Lili Lu. 2018. Primary investigation of the implementation of "dynamic + broad consent" in medical institutions (original in Chinese: "动态 + 泛知情同意"在医疗机构实施初探). *Chinese Medical Ethics* 31:487–491.

Chen, Yan, Deepanwita Banerjee, Aindrila Mukhopadhyay, and Christopher J. Petzold. 2020. Systems and synthetic biology tools for advanced bioproduction hosts. *Current Opinion in Biotechnology* 64:101–109.

Chen, Yongxi, and Lingqiao Song. 2018. China: Concurring regulation of cross-border genomic data sharing for statist control and individual protection. *Human Genetics* 137 (8): 605–615.

Chu, Chengjun, and Jun Liu. 2002. Transnational protection of human gene resource extraction: A case study based on Harvard University's gene research project in Anhui Province (original in Chinese: 人类基因资源提取的跨国保护——基于哈佛大学在安徽的基因研究项目的个案分析). *Law Science* 11:60–66.

Conroy, Megan, Jonathan Sellors, Mark Effingham, Thomas J Littlejohns, Chris Boultwood, Lorraine Gillions, CLM Sudlow, Rory Collins, and Naomi E Allen. 2019. "The advantages of UK Biobank's open-access strategy for health research." Journal of Internal Medicine 286 (4): 389–397.

Department for Digital, Culture, Media & Sport of the United Kimdom (DCMS). 2021. *Data: A new direction*. Accessed 15 March 2026. https://www.gov.uk/government/consultations/data-a-new-direction.

Department for Digital, Culture, Media & Sport of the United Kimdom (DCMS). 2022. *Consultation outcome–Data: A new direction–government response to consultation*. Accessed 15 March 2026. https://www.gov.uk/government/consultations/data-a-new-direction/outcome/data-a-new-direction-government-response-to-consultation#ch1.

Department of Health and Social Care, Department for Business, Energy & Industrial Strategy, Office for Life Sciences, and Lord Bethell of Romford. 2020. *Genome UK: The future of healthcare*. Accessed 15 March 2026. https://www.gov.uk/government/publications/genome-uk-the-future-of-healthcare.

Eckstein, Lisa, Donald Chalmers, Christine Critchley, Ruthie Jeanneret, Rebekah McWhirter, Jane Nielsen, Margaret Otlowski, and Dianne Nicol. 2018. Australia: Regulating genomic data sharing to promote public trust. *Human Genetics* 137 (8): 583–591.

Freemont, Paul S. 2019. Synthetic biology industry: Data-driven design is creating new opportunities in biotechnology. *Emerging Topics in Life Sciences* 3 (5): 651–657.

Global Biofoundry Alliance. 2023. *About the GBA?*. Accessed 15 March 2026. https://www.biofoundries.org/about.

Helgesson, Gert. 2012. In defense of broad consent. *Cambridge Quarterly of Healthcare Ethics* 21 (1): 40–50.

Huang, Zhixiong, and Xinyu Wei. 2021. The rule games between the US and EU on transborder data flow and China's response: From the persepctive of invalidation judgment of the privacy shield framework (original in Chinese: 美欧跨境数据流动规则博弈及中国因应——以《隐私盾协议》无效判决为视角). *Journal of Tongji University (Social Science section)* 32:31–43.

Kaye, Jane, Liam Curren, Nick Anderson, Kelly Edwards, Stephanie M Fullerton, Nadja Kanellopoulou, David Lund, Daniel G MacArthur, Deborah Mascalzoni, and James Shepherd. 2012. "From patients to partners: Participant-centric initiatives in biomedical research." Nature Reviews Genetics 13 (5): 371–376.

Kaye, Jane, Edgar A. Whitley, David Lund, Michael Morrison, Harriet Teare, and Karen Melham. 2015. Dynamic consent: A patient interface for twenty-first century research networks. *European Journal of Human Genetics* 23 (2): 141–146.

Knoppers, Bartha Maria. 2014. Framework for responsible sharing of genomic and health-related data. *The HUGO Journal* 8 (1): 3.

Lewis, Dyani. 2023. *China's souped-up data privacy laws deter researchers*. Nature. Accessed 15 March 2026. https://www.nature.com/articles/d41586-023-01638-1.

Li, Xiaojie, Yali Cong, and Ruishuang Liu. 2022. Research under China's personal information law. *Science* 378 (6621): 713–715.

Ma, Qijia, and Feihu Liu. 2022. Discussion on national security governance in cross-border data transfer (original in Chinese: 数据出境中的国家安全治理探讨). *Theoretical Exploration*105–113.

Mallapaty, Smriti. 2022. China expands control over genetic data used in scientific research. *Nature* 605 (7910): 405–405.

Ministry of Science and Technology of the People's Republic of China (MOST). 2022a. *Notice on updating the FAQ on human genetic resources management* (Original in Chinese: 关于更新人类遗传资源管理常见问题解答的通知). Accessed 15 March 2026. https://www.most.gov.cn/tztg/202203/t20220304_179634.html

Ministry of Science and Technology of the People's Republic of China (MOST). 2022b. *Notice on updating the FAQ on human genetic resources management* (Series 2), (Original in Chinese: 关于更新人类遗传资源管理常见问题解答(系列问答二)的通知). Accessed 15 March 2026. https://www.most.gov.cn/tztg/202204/t20220415_180263.html.

Mu, Tong, and Wenxuan Yu. 2021. On the optimization path for legal mechanism for benefit-sharing to biological genetic resources (original in Chinese: 我国生物遗传资源惠益分享法律机制的优化路径). *Journal of Ecology and Rural Environment* 37:1104–1108.

National Computer Network Emergency Response Technical Team/Coordination Center of China (CNCERT). 2021. *China internet network security annual report 2020* (Original in Chinese:

2020年中国互联网网络安全报告) ". Accessed 15 March 2026. https://www.cert.org.cn/publish/main/46/2021/20210721130944504525772/20210721130944504525772_.html.

National Development and Reform Commission of People's Republic of China (NDRC). 2021. *Notice of the national development and reform commission on circulating the 14th Five-Year Plan for Bio-Economic Development*, translated by authors of this paper (Original in Chinese: 国家发展改革委关于印发《"十四五"生物经济发展规划》的通知). Accessed 15 March 2026. https://www.ndrc.gov.cn/xxgk/zcfb/ghwb/202205/t20220510_1324436.html

National Health Commission of People's Republic of China (NHC). 2021. *Announcement on public comments on measures for ethical review of life sciences and medical research involving humans* (Draft for Comment) (Original in Chinese:关于涉及人的生命科学和医学研究伦理审查办法(征求意见稿)公开征求意见的公告)." Accessed 15 March 2026. https://www.nhc.gov.cn/qjjys/c100016/202103/fab2b35526c8490497f401630b55a2b5.shtml.

National Health Commission of People's Republic of China (NHC). 2025. 人类遗传资源管理专题专栏 (Special Topic on Human Genetic Resource Management, translated by authors of this paper). Accessed 15 March 2026. https://www.nhc.gov.cn/qjjys/rlyczygl/new_list.shtml

Ouyang, Ting, Yinfeng Yang, Jianhua Shu, Lei Zhang, and Dan Wang. 2020. Ethical thinking on patients' awareness of privacy protection in the context of health big data (original in Chinese: 健康大数据背景下患者隐私保护意识调查及伦理思考). *Journal of Chaohu University* 22:91–97.

Peng, Changhui, Xinzhang Song, Hong Jiang, Qiuan Zhu, Huai Chen, Jing M Chen, Peng Gong, Chang Jie, Wenhua Xiang, and Guirui Yu. 2016. "Towards a paradigm for open and free sharing of scientific data on global change science in China." Ecosystem Health and Sustainability 2 (5): e01225.

Prictor, Megan, A. Megan, Ainsley J. Newson Lewis, Matilda Haas, Sachiko Baba, Hannah Kim, Minori Kokado, Jusaku Minari, Fruzsina Molnar-Gabor, and Beverley Yamamoto. 2020. Dynamic consent: An evaluation and reporting framework. *Journal of Empirical Research on Human Research Ethics* 15 (3): 175–186.

Pu, Ye. 2018. *Six penalties sent a signal for the protection of human genetic resources*, (Original in Chinese: 六份处罚书发出人类遗传资源保护信号弹). Accessed 15 March 2026. https://news.sciencenet.cn/htmlnews/2018/10/419337.shtm

Sharma, Yojana. 2022. *New genetic research rules: What impact on collaboration?*. Accessed 15 March 2026. https://www.universityworldnews.com/post.php?story=20220407145532404

Steinsbekk, Kristin Solum, Bjørn Kåre Myskja, and Berge Solberg. 2013. Broad consent versus dynamic consent in biobank research: Is passive participation an ethical problem? *European Journal of Human Genetics* 21 (9): 897–902.

Sudlow, Cathie, John Gallacher, Naomi Allen, Valerie Beral, Paul Burton, John Danesh, Paul Downey, Paul Elliott, Jane Green, and Martin Landray. 2015. UK biobank: An open access resource for identifying the causes of a wide range of complex diseases of middle and old age. *PLoS Medicine* 12 (3): e1001779.

Szustakowski, Joseph D., Suganthi Balasubramanian, Erika Kvikstad, Shareef Khalid, G. Paola, Ariella Sasson Bronson, Emily Wong, Daren Liu, J. Wade Davis, and Carolina Haefliger. 2021. Advancing human genetics research and drug discovery through exome sequencing of the UK biobank. *Nature Genetics* 53 (7): 942–948.

Tou, Xiaodong, and Xiurong Huang. 2015. When the "modern interest game" turns to the "traditional field of genetic resource": The research on the deep root of the imbalance of the benefit distribution of intellectual property rights of genetic resources and its correction principle (original in Chinese: 当"现代的利益博弈"转向"传统的遗传资源领域"——遗传资源知识产权惠益分配失衡的深层根源及其矫正原理研究). *Research on Rule of Law* 2:142–151.

Wörheide, Maria A., Jan Krumsiek, Gabi Kastenmüller, and Matthias Arnold. 2021. Multi-omics integration in biomedical research–a metabolomics-centric review. *Analytica Chimica Acta* 1141:144–162.

Wu, Jiarui. 2022. Seeking the balance between security protection and open sharing of life and health big data–Reflections on the Personal Information Protection Law of the People's

Republic of China (Original in Chinese: 寻找生命健康大数据在安全保护与开放共享之间的平衡—对《中华人民共和国个人信息保护法》的思考). *Chinese Bulletin of Life Sciences* 34:1–4.

Yang, Heqing. 2022. Rules on cross-border provision of personal information (original in Chinese: 个人信息跨境提供的规则). In *Interpretation of the personal information protection law of the People's Republic of China (original in Chinese: 中华人民共和国个人信息保护法释义)*, edited by Heqing Yang, 99–111. Beijing: Legal Press.

Yu, Xiang, and Wei Li. 2014. Informed consent and ethical review in Chinese human experimentation: Reflections on the "Golden Rice event". *Biotechnology Law Report* 33 (4): 155–160.

Zhang, Chenglin, Xiaoyun Chen, and Minsheng Fan. 2017. Domestic and foreign application of extensive informed consent in future research (original in Chinese: 国内外广泛知情同意应用于"将来研究"的探讨). *Chinese Medical Ethics* 30:760–764.

Zhang, Taolue, and Huizhao Tu. 2021. Legal regulation of cross-border flow of data of intelligent networked vehicles (original in Chinese: 智能网联汽车数据跨境流动的法律规制). *Dispute Settlement* 7:98–104.

Zhang, Joy Y. 2011. The 'National'and the 'cosmos' the emergence of synthetic biology in China. *EMBO Reports* 12 (4): 302–306.

Zhang, Joy Y. 2022. Updating the understanding of data protection and promoting international cooperation structurally (Original in Chinese: 更新对数据保护的认知，从结构上推动国际合作). *Science and Technology of China* 9:22–25.

Zhang, Joy Y., Sonia Ben Ouagrham-Gormley, M. Kathleen, and Vogel. 2022. Creating common ground with Chinese researchers. *Issues in Science and Technology* 38 (4): 45–48.

Zhang, Joy Y., Sonia Ben Ouagrham-Gormley, M. Kathleen, and Vogel. 2023. Chinese academics are becoming a force for good governance. *Issues in Science and Technology* 39 (4): 81–84.

Zhao, Jingwu. 2023. On the systematization of data cross-border assessment, contracts and authentication rules (original in Chinese 论数据出境评估、合同与认证规则的体系化). *Administrative Law Review* 1:78–94.

Zhangyu Wang is a PhD candidate at the Faculty of Law, University of Macau. His research interest covers legal issues related to cross-border health data transfer and data protection, medical tourism, and the governance of emerging biotechnologies.

Li Du is an Associate Professor at the Faculty of Law and a Research Fellow at the Asia-Pacific Academy of Economics and Management at the University of Macau, Macau SAR, where he also serves as Director of the Centre for Empirical Legal Studies and PhD Program Coordinator. He holds dual bachelor's degrees in clinical medicine and law from Wuhan University, China, and a PhD in law from the University of Alberta, Canada. This unique interdisciplinary background in both medicine and law informs his research at the intersection of law, science, and technology. His teaching and research interests span international law, food law, biotechnology law and policy, and data protection law. He has led numerous research projects examining the legal and ethical implications of novel and emerging technologies, including synthetic biology, AI-enabled biomedical research, and bio-agricultural products. Adopting comparative and interdisciplinary approaches, his work seeks to understand how regulatory frameworks can adapt to rapid scientific advancements while balancing innovation with safety, ethics, and public trust. Du frequently presents his research at international conferences and publishes regularly in leading academic journals. He is actively engaged in scholarly networks focused on the responsible governance of emerging technologies and contributes to relevant policy discussions.

Chapter 6
China's Role in Global Risk Governance: A Comparative Legal Perspective

Simona Novaretti

Abstract Since the Opium Wars, China has been characterised as the 'Oriental Other': 'without God or law', according to Marcel Granet's celebrated 1934 assertion. Contemporary discourse has extended this characterisation, with China often portrayed as lacking morality or possessing a questionable moral framework—a perception that recent controversies concerning clinical practices and biotechnological applications in the life sciences within the PRC appear, superficially, to substantiate.

This chapter examines China's evolving legal identity and its implications for global biogovernance. Drawing upon China's historical role as a recipient of externally imposed legal models over the past two centuries, the analysis interrogates two fundamental questions: first, whether a distinctive 'Chinese legal model' exists, rooted both in foreign (particularly European) legal frameworks and indigenous Chinese values of traditional and ideological origins; secondly, how these characteristics shape contemporary Chinese legal approaches to biogovernance, and whether such approaches align with, contradict, or provide novel perspectives compared to Western legal traditions. Ultimately, it assesses how China—now a major global scientific and economic actor—can meaningfully contribute to the global commons, moving beyond orientalist assumptions to engage with the complexities of China's legal and regulatory reality in the life sciences.

China: A Country 'Without'

In recent years, bioscience has played a major role in the People's Republic of China (hereinafter: PRC)'s global competitiveness. PRC's 13th Five-Year Plan (2016–2020) designated biomedicine, synthetic biology, and regenerative medicine

S. Novaretti (✉)
Department of Law, University of Turin, Turin, Italy
e-mail: simona.novaretti@unito.it

J. Y. Zhang et al. (eds.), *BioGovernance Commons and China's Role in Global Risk Governance*, Emerging Globalities and Civilizational Perspectives,
https://doi.org/10.1007/978-3-032-32198-5_6

as strategic fields and industries (Mallapaty 2018). Consequently, by 2016, China was already being highlighted as a 'rising star' in the Nature Index, an open database of author affiliations and institutional relationships, with numerous Chinese institutions demonstrating rapid improvement in their research output and ranking (Woolston 2023). In 2022, China overtook the United States in natural sciences for the first time, and in 2024, China advanced from second to first place for its overall contribution to research in the four natural science categories covered by the same database: physical sciences, chemistry, biological sciences, and Earth and environmental sciences (Nature Index 2025).

The international community follows China's growing pursuit of technological proficiency with a mixture of admiration and concern, with the latter usually prevailing. Articles in international publications describing China's accomplishments frequently voice concerns that China could leverage its technical monopoly to impose its authoritarian political model on the rest of the globe. Additionally, many foreign observers depict the PRC as the 'Wild East' of biomedical research, pointing to the country's allegedly 'lax ethical environment' and purported lack of proper legislation as the principal reasons behind its rise in medicine (Tatlow 2015).

In 1934, Marcel Granet described to the Western public the essence of China as having 'neither God, nor Law' (Granet 1934). Should we conclude that it has no ethics, especially bioethics, either?

At least from the standpoint of the PRC's written laws, this does not appear to be the case. On 17 October 2020, the Standing Committee of the PRC's National People's Congress passed China's first national law dedicated to biosecurity and biosafety, the 'Biosecurity Law' (*Shenwu Anquan fa*生物安全法) of the PRC. In the law, the aim to 'maintain national security, prevent and respond to biosecurity hazards, safeguard the lives and health of people, protect biological resources and the environment' is complemented by the purposes to promote 'the sound development of biotechnology' and to foster 'the construction of a community with the shared future for mankind and the harmonious coexistence of man and nature.' Moreover, many Chinese statutes enacted or amended since the beginning of the second decade of the twenty-first century contain references to ethics, morality, or related principles, particularly when scientific research or medical practice is at stake. We can take as an example Article 1009 of the PRC Civil Code (*Minfadian*民法典, 2020): 'Medical and scientific research activities concerning human genes and human embryos, among others, shall be carried out in accordance with the laws, administrative regulations, and relevant provisions issued by the state, without endangering human health, violating moral principles, or damaging public interests.' Similarly, Article 32 of the Law of the People's Republic of China on the Promotion of Basic Medical and Health Care (*Jiben yiliao weisheng yu jiakang cujin fa* 基本医疗卫生与健康促进法, 2019) stipulates: 'Clinical experiments and other medical research of drugs and medical devices shall comply with the medical ethics and specifications, pass ethical review according to the law, and obtain informed consent'; Article 20 of the Medicinal Product Administration Law of the People's Republic of China (*Yaoping guanli* 药品管理法, adopted in 1984, revised in 2001, 2013, 2015, and 2019) further provides Pharmaceutical clinical trials shall be conducted in conformity with ethical principles, and clinical trial protocols shall be developed, subject to the review and consent of the ethics committee'; Article 51 of the Scientific and Technological Progress Law of the People's Republic of China (*Kexue jishu jinbu fa* 科学技术进步法, adopted in 1993 and revised in 2007 and 2021) mandates: 'Scientific and technological research and development institutions shall formulate their bylaws/internal regulations in accordance with the law, and conduct scientific and technological research and development activities in accordance with the functional positioning and business scope stipulated therein; they shall strengthen the improvement of conduct in scientific research and studies, establish and improve scientific research integrity and scientific and technological ethics management systems, and comply with good scientific research practices; they shall not organise, participate in, or support superstitious activities' (Novaretti 2025).

All these provisions establish a duty for researchers and research institutes to comply with ethical standards and implement control mechanisms to be carried out by review and supervisory committees. The importance attached to ethics in the current Chinese context led the legislature in 2023 to include a reference to it in the Law on Legislation (*Lifa fa*立法法), Article 8 of which now reads: 'Legislation shall [...] adhere to the combination of the rule of law and the rule of virtue, forge a strong sense of community of the Chinese nation, and promote socialist cultural-ethical progress'.

Hence, why so much suspicion about China's ethical conception and the importance it attributes to it? Is it merely a problem of application or ineffectiveness of these provisions, or is there something intrinsically problematic in the Chinese legal system?

To be honest, life science is only one of many fields in which the PRC is perceived as a 'global risk'. From the perspective of comparative law, the country has historically been a recipient of externally imposed legal models, which raises fundamental questions about the nature and origin of its legal system. Does a distinctive 'Chinese legal model' exist, and if so, what are its roots? Can these 'Chinese characteristics' (*Zhongguo tese* 中国特色) shape a contemporary Chinese legal approach to biogovernance? Do they align with, contradict, or provide novel perspectives compared to international legislation and regulations within the Western legal tradition?

China as a 'Country Without Law'

'China Has No Law'...but Why? China as a Mirror of Western Expectations

Almost a century after the publication of Granet's book on 'Chinese Thought', the common perception of the Chinese legal system has not changed substantially, either amongst ordinary people or academics. There is no need to cite newspaper articles or statements by political leaders. Like many other Chinese law scholars, I have lost track of how many times—after mentioning my field of expertise—I have been asked the same questions: 'Do they really have law in China?' or even 'Chinese Law is an oxymoron, isn't it?' This occurs even when the interlocutor is a legal comparatist or a sinologist. Indeed, comparative law scholars specialising in Chinese law are too often considered by their colleagues to have limited scientific interests, regardless of how many different areas of Chinese law they have researched and written about. The root of this attitude can partly be traced back to Granet's prejudice regarding China's lack of law: at the end of the day, what could be more academically limited (and, therefore, insignificant) than studying a non-existent legal system?

Such misbelief is deeply ingrained and globally widespread. In fact, comments on the non-existence of Chinese law prompted Teemu Ruskola, at the beginning of the twenty-first century, to investigate the origins of such a misconception and why it easily gains traction with so many people (Ruskola 2002). Eventually, drawing inspiration from the title of Edward Said's famous monograph on 'Orientalism', he formulated the notion of 'legal Orientalism', defined as 'a set of interlocking narratives about what is and what is not law, and who are and who are not its proper subjects' (Ruskola 2013).

In the past three decades, Ruskola's approach has been successfully employed to analyse the West's representations of East Asia, especially China. Once again, through these depictions, the West expresses a judgement on itself: China was what the West was not (again: 'non-Europe') in a positive or in a negative way, depending on historical moments and authors. For example, during the seventeenth and eighteenth centuries, Europeans admired China and generally considered it superior to other countries in administration, politics, and order. At a time when religious wars ravaged Europe, Jesuits such as Matteo Ricci and thinkers such as Voltaire praised Chinese stability and religious tolerance, and extolled its government, run in accordance with Confucian principles, with only the most educated and virtuous men eligible for state office (Ruskola 2002). Even the Chinese language was greatly appreciated. Famous, in this regard, is the opinion of Leibniz, who regarded the Chinese script as the (albeit only partial) realisation of his ideal of a universal language (Roy 1972). However, at the beginning of the nineteenth century, the European situation changed and, with it, Europeans' attitude towards China also changed. China's extraordinary stability and immobility, hitherto considered strengths, became its principal flaws. Amongst the most striking and most cited examples of this paradigm shift is surely Hegel's 'Philosophy of History', in which the German philosopher stated that 'The History of the World travels from East to West, for Europe is absolutely the end of History, Asia the beginning' (Hegel 2001). China's main characteristics are durability and stability: for China, 'every change is excluded'. Therefore, it lies outside History, in a sort of 'unhistorical History', which represents 'the childhood' of human historical development, and the paradigmatic example of 'Oriental Despotism' (Hegel 2001). Similar criticism of China and of the immutability and stability of its culture and institutions, especially when compared with the West's attitude to movement and development, can be found in the writings of Marx and Weber. Marx's description of China as 'vegetating in the teeth of time' is well-known (Marx 1858), as is Weber's opinion on the static nature of Chinese intellectual life, to which he attributed its failure to develop modern capitalism (Weber 1968).

In sum, throughout European modern history, China was used as a mirror, capable of reflecting the image that European thinkers wished belonged, or did not belong, to the West, or (more frequently) demonstrating how far more advanced European culture was (Ruskola 2002). In this sense, as stated by Derrida with reference to the Chinese system of writing, the image of China was a sort of European 'hallucination', which 'translated less an ignorance than a misunderstanding' (Derrida 2016).

China Has No Law…from When? The Nineteenth-Century Encounter

It was in this historical, philosophical, and cultural context that the first (real) encounter between European and Chinese law took place. The European jurists who arrived in China after the Chinese defeat in the Opium Wars sought in the Chinese legal system the elements to which they were accustomed, i.e., what, in nineteenth-century Europe, was considered to be 'The' law. For continental European jurists, who were living the era of codification and steeped in legal positivism, the hallmark of a modern legal system was the presence of a civil code or (at least) a well-developed system of civil law rules sustained by sound jurisprudence. English jurists, on the other hand, considered it essential to engage with legal professionals who were trained and selected based on specific technical knowledge and who followed fixed, rigid procedures with predictable outcomes based on judicial precedents. What they found in China was, instead, a system based on the idea of 'social harmony', in which legal norms were conceived as instruments in the hands of the sovereign and his officials, to be used to govern the people and punish those who had acted contrary to their duty, proving resistant to all teaching and advice from their superiors in age or rank (Chen 2015). The areas that had formed the core of Roman law development, and around which the first nineteenth-century codes had been drafted, namely property and contracts, were hardly taken into consideration by imperial legislation, which considered them, at least rhetorically, to be 'minor matters', the decision on which could be delegated to the relevant social groups—family, clan, village, or trade guild (Jones et al. 1994). Instead, Chinese Imperial codes focused mostly on administrative or criminal provisions, with the latter including extremely cruel punishments for the most serious crimes. These provisions were kept unknown by the common people, in accordance with the Confucian idea that knowledge of the law would corrupt their minds, leading subjects to seek ways to circumvent legal norms instead of focusing on their *li* 礼, generally translated as 'ritual', but here used in the sense of the conduct appropriate to one's social and/or family status (Head and Wang 2005). A system in which suffering injustice was considered preferable to going to court, since all parties involved in a dispute were held responsible for disrupting social harmony, was certainly not conducive to the development of the legal professions. It is no coincidence that when, between the nineteenth and twentieth centuries, Chinese scholars in charge of the reforms had to translate the concept of 'lawyer' into Chinese, they decided to invent a new expression, *lüshi* 律师 (lit. 'Master of the law') instead of using the terms which, until that time, identified those who assisted the parties in court proceedings. These were called *songshi* 讼师 (lit.: litigation master) or even *songgun* 讼棍 (lit.: litigation trickster) and were often viewed with suspicion, since they were considered to take advantage of social disorder, if not to cause it (Chen 2024). Even more shocking than the lack of 'true' lawyers and legal scholars was probably the fact that not even Imperial magistrates were 'properly' (according to Western standards) trained in law. Instead, they had often spent years, when not decades, studying the literary, philosophical, and poetic texts that constituted the corpus of the ideology on which the state was founded: Confucianism. It was the knowledge of the latter that was tested by the state examinations necessary to enter the ranks of the imperial administration, to which the magistrates belonged (Wang 2024).

No aspect of this system resembled 'law' in the eyes of the Western invaders. Thus, a legal system that for millennia had governed a territory larger than that governed by Roman law not only during the imperial age, but throughout the Middle Ages and up to the modern age (Jones et al. 1994), and which, according to some calculations, had produced the highest GDP in the world for many centuries (Chen and Peng 2022), was considered non-existent and forced to reform, adapting to the legal standards and values imposed by the victors.

'China Has No Law'…but It Will Change: Qing and Nationalist Reforms (1903–1949)

During the early decades of what the Chinese call the 'century of humiliation' (1839–1949), and more precisely until the end of the nineteenth century, the Chinese court resisted both external pressures from the foreign powers, which were pushing for a Westernisation of the system, and recommendations from part of the country's intellectual elite, in particular the members of the Yangwu Movement (*Yangwu yundong*洋务运动, also known as *ziqiang yundong*自强运, litt. Self-strengthen movement), who suggested the adoption of Western techniques and theories. In the latter's view, the invaders' skills could be useful to make China great again, according to the motto 'Chinese knowledge as a foundation, Western knowledge as a means'(*zhongxue wei ti, xixue wei yong* 中学为体西学为用. However, it was only after the humiliation suffered by the defeat by Japan in 1894–1895, coupled with the repeated demands of extraterritoriality by outside powers from 1895 to 1902, which became even greater as a result of the Boxer Revolution of 1900, that the need to implement reforms to Westernise the country became urgent and unpostponable. These reforms, started in the last years of the Qing Dynasty, were only to be completed 30 years later with the enactment of the Six Codes of the Guomindang (国民党, lit. Nationalist Party, hereinafter: GMD) by the Nationalist government (Li 2022b).

It is worth mentioning that in the late Qing era and in the early years of the Republic of China, the legal model chosen, particularly in the civil sphere, was the German one, filtered through the Japanese experience. This choice was never really abandoned, despite the drastic political and ideological changes experienced by the country in the twentieth century, which from time to time also implied the inclusion of institutions and legal concepts of different origins. In particular, as early as 1927, with the conquest of Nanjing by the GMD forces, the goal was no longer simply the Westernisation of law, but a real modernisation of the legal system. Nationalist legal scholars aimed to synthesise the latest developments in international legal theory, the characteristics of Chinese tradition, and the needs of the 'new China' (Huang 2001). To this end, they made extensive use of comparative law, and by the early 1930s, they succeeded in providing the Republic of China with a 'modern' and comprehensive body of law. Their efforts, although appreciated internationally, failed to shake China's image as a lawless country. Not only were the laws in question—enacted by a power that controlled little more than half of China's provinces—not in force throughout China, but even in the areas controlled by the GMD, their application was relatively limited. The struggle between the newly formed Chinese Communist Party and the GMD at the turn between the 1920s and 1930s, the Japanese invasion of 1939, the Second World War, and, finally, the 'war of liberation' between Maoist and GMD forces (1946–1949) prevented the new legal rules from truly taking hold outside the large port cities, which were more open to the profound social changes brought about, for example, by the gender equality established by the new rules on marriage, filiation, and succession (Huang 2001). These rules were then formally eliminated after the Communist victory. From that moment on, and for almost 30 years, China was truly a lawless country.

China Has No Law (Again!): From the Maoist Era (1949–1978) to the Enactment of the Civil Code of the People's Republic of China (2020)

On 1 October 1949, the People's Republic of China (hereinafter: PRC) was founded. Two days earlier, the First Plenary Session of the Chinese People's Political Consultative Conference had adopted the Common Programme, the first (provisional) constitution of the PRC, which abolished, in Article 17, 'all laws, decrees and judicial systems of the Guomindang's reactionary government which oppressed the people'. Despite the same article promising that 'laws and decrees protecting the people shall be enacted, and the people's judicial system shall be established', such outcomes were never achieved during Maoism. It was only in 1978, with the beginning of the reform and opening-up period, that China began the process that would lead, in March 2011, to Wu Banguo, at that time Chairman of the Standing Committee of the National People's Congress, declaring that the PRC had a 'complete socialist legal system with Chinese characteristics' (Information Office of the State Council 2011). It is worth noting that, whilst in the early 1980s it was especially the 1950s' Soviet-inspired legislation that was taken as reference, as the reform progressed, the system became increasingly comprehensive and sophisticated, thanks also to exchanges and legal training programmes between Chinese public institutions, government bodies, and universities and their foreign counterparts, particularly European (especially German) and American ones (Baskir 2015). However, the Chinese legal system still lacked the norm that has been considered the hallmark of legal development in a civil law country and the most mature product of its doctrine since the enactment of the Napoleonic Code in France in 1804: the Civil Code. In this respect, too, the differences between Chinese law and that of countries belonging to the Roman-Germanic legal tradition were eventually eliminated in 2020, with the enactment of the first Civil Code of the PRC.

Like every other legal product resulting from the reforms, this event has not substantially changed prejudices regarding the value of law in the PRC. In fact, whilst nobody can now seriously affirm that China is a lawless country, it can still be argued that its laws are irrelevant, and somebody has gone so far as to even question the possibility of using 'Western' legal terminology to convey the concepts of that system (Clarke 2020).

I have had the opportunity to refute these claims elsewhere (Novaretti 2025). Here, I think it is more interesting to focus on some characteristics of the Civil Code which, at least in my opinion, can illustrate better than others the possible contributions of the PRC to the biogovernance commons from the perspective of comparative private law.

The Chinese Civil Code, Between Foreign Legal Models and Original Inputs

Foreign Legal Models

The Civil Code of the People's Republic of China was approved on 28th May 2020. The code comprises 1260 articles organised into seven books: General Part, Real Rights, Contracts, Personality Rights, Marriage and Family, Succession, and Tort Liability. Whilst Prof. Wang Liming claimed in 2019 that the promulgation of the Chinese civil code would constitute 'a meaningful and momentous chapter in the history of civil law', comparable to the French Code of 1804 and the German Code of 1900, such assessments remain contested. In fact, the Chinese Civil Code is not generally considered to be a fully original norm; in particular, its structure and content are still profoundly influenced by the German model, albeit modernised and supplemented by solutions from other 'Western' legal systems (Hao and Sirena 2023). This is not to say that it does not contain original solutions, which were incorporated into the Chinese legal system, both traditional and socialist values. As an example, we can mention Art. 1186 of CC, according to which 'If the damage is through no fault of both the victim and the tortfeasor, the loss may be borne by both parties according to the law'. This provision—which leaves the decision whether to attribute liability to the discretion of the judge, who will assess the ability of the parties to bear the costs—contrasts with Western concepts of liability and the idea that private disputes are an area of commutative justice. The article draws upon the Soviet model, in particular Article 406 of the Russian Civil Code of 1922, which, however, was little used in Russia. It also reflects the traditional Chinese conception that the strong must help the weak, revealing the concern for the distribution of wealth, and therefore the focus on distributive justice, which is a characteristic of the Chinese legal system (Hao 2021). Other 'Chinese characteristic' values are included in Book 1 of the Civil Code, which is dedicated to the General Part. It contains articles that refer to both the traditional Chinese values and values more recently embraced by the Chinese leadership regarding sustainable development. Art. 9, according to which 'Civil entities engaging in civil activities shall benefit conservation of resources and protection of the ecology and environment', is a clear example of the latter, whilst the traditional value of filial piety is recalled by Article 26(2): 'Adult children have an obligation to provide for, support, and protect their parents', and that of 'benevolence/humanity' by Article 184: 'When voluntarily carrying out emergency assistance causes harm to the person receiving assistance, the aider does not bear civil liability' (Novaretti 2019). However, it is especially Chap. V of Book 1, which addresses Civil Rights, and Book 7, dedicated to 'Personality Rights', that represent respectively the most innovative and original aspects of the PRC's Civil Code.

Innovation and 'Original' Inputs

As just recalled, Book 1, Chap. V of the Civil Code addresses civil rights. In particular, Articles 109–110 represent features of a 'modern' civil code, listing inalienable rights not traditionally mentioned in civil provisions. These are personal liberty and personal/human dignity (*renge zunyan*人格尊严), the right to life, the right to corporeal integrity, the right to health, the right to name, the right to likeness, the right to reputation, the right to honour, the right to privacy, and the right to freedom of marriage. With reference to the right to privacy, Article 111 is particularly significant: 'The personal information of a natural person shall be protected by law. Any organisation or individual needing to obtain the personal information of other persons shall legally obtain and ensure the security of such information, and shall not illegally collect, use, process, or transmit the personal information of other persons, nor illegally buy, sell, provide, or publish the personal information of other persons'.

This approach may be compared with the Quebec Civil Code (1991), whose Preliminary Provisions state: 'The Civil Code of Quebec, in harmony with the Charter of the French language (chapter C-11), the Charter of human rights and freedoms (chapter C-12) and the general principles of law, governs persons, relations between persons, and property'. Article 3 of the same Code then provides: 'Every person is the holder of personality rights, such as the right to life, the right to the inviolability and integrity of his person, and the right to the respect of his name, reputation and privacy. These rights are inalienable'.

The most original part of the Chinese Civil Code is undoubtedly Book 7, dedicated to personality rights (Hao 2021). Book 7 not only lists and seeks to protect the personality rights enumerated in Chap. V, Book 1, but also aims to safeguard all other rights which pertain to being a person as such. Article 990 of the Civil Code provides: 'Personality rights shall refer to the rights of life, body, health, name, image, reputation, honour, and privacy, among others, enjoyed by parties to civil legal relations. Besides the personality rights prescribed in the preceding paragraph, a natural person enjoys other personality rights based on personal freedom and personal dignity (*renge zunyan*人格尊严)'.

Some authors have criticised the decision to include Book 7 into the Code, considering it a useless complication (Hao 2021). On the contrary, according to other authors, the legislators' choice was a response both to the growing consciousness of personality rights amongst the Chinese public and to the ubiquitous technological risks in contemporary China. Article 990 in particular could be used to protect constitutional rights, representing a way to adapt the law to respond to new biotechnologies (Zhou 周 2020).

Amongst the values listed in it, the principle of human dignity seems to me particularly important, given its relevance both in the most recent bioethical debate at the international level and in Chinese traditional culture. Henceforth, the remainder of this chapter shall be devoted to this fundamental principle. I will start by analysing its role within international treaty law and soft law dedicated to the regulation of biotechnology. This was formulated by leading international organisations mainly as a result of Western philosophical reflections and replicated in national (Western) legislation at various levels. I will then move on to the meaning attributed to the concept in the Chinese ethical-philosophical context. This will enable me to demonstrate how its incorporation into the primary civil law of the PRC could offer fresh insights into global risk governance.

Human Dignity: A 'Western' Legal Concept, with 'Chinese Characteristics'

Human Dignity in International Treaties and Soft Law and in (Western) National Legislation

Recent technological advances in genetic engineering have enabled intervention in previously inaccessible aspects of human biology. While offering hope for treating incurable diseases, these developments have prompted the scientific community and public opinion to question the boundaries between possibility and appropriateness, particularly regarding human germline genetic editing (HGGE)—the manipulation of reproductive cells or embryos capable of transmitting changes to future generations.

The need to redefine 'natural' and 'human' boundaries has reinvigorated bioethical reflection, exposing the relativity and internal contradictions of seemingly self-evident principles (Albers 2018, 510). Central among these is the concept of 'human dignity', which inevitably becomes contested when the meaning of 'human' itself is under discussion. In contemporary bioethical debate, human dignity is invoked both to support individual self-determination and to limit individual autonomy for protecting personal and species-level dignity.

These positions carry significant practical and legal implications. Since 1948 and the UN Universal Declaration of Human Rights, human dignity has occupied a pivotal role in interpreting internationally recognised human rights, reaffirmed in numerous regional and global instruments and enshrined in many national constitutions across legal traditions (Brownsword 2014, 1). However, domestic law reveals profound tensions between liberal interpretations—defending individual choice—and conservative approaches viewing dignity as a constraint on individual autonomy, protecting persons even from consensual abuses.

The French Council of State's 1995 ruling (No. 136727) prohibiting 'dwarf tossing', even with the participants' consent, is an example of a conservative approach. More relevant to biotechnology, the 1997 Oviedo Convention on Human Rights and Biomedicine permits somatic gene therapy while absolutely prohibiting HGGE. Article 13 allows genome modification only 'for preventive, diagnostic or therapeutic purposes' and excludes interventions 'intended to introduce a modification in the genome of descendants.' The explanatory report justifies this prohibition by citing risks to 'not only the individual, but the species as a whole,' particularly preventing intentional modifications producing 'individuals or entire groups with particular characteristics and qualities' (para. 89).

Though not explicitly mentioned in Article 13's commentary, human dignity underpins both the Convention generally (Andorno 2009, 5) and this provision specifically, as evidenced by the Preamble's recognition of the need 'to respect human beings both as individuals and as members of the human species' and Article 1's commitment to protect 'human beings in their dignity and identity' (van Beers 2020, 12). This interpretation extending 'human' beyond individuals to encompass humanity and future generations reflects the Council of Europe's conservative bioethical stance. Despite being the first binding international biotechnology instrument, the Convention remains unratified by several signatories nearly 30 years later, while four EU members—Germany, Austria, Belgium, and Malta—have not signed it, revealing the persistent divergence even within culturally homogeneous regions.

In light of this, it is legitimate to ask how the discourse on human dignity is articulated, and what impact this has on the regulation of biotechnology within a context that is traditionally perceived as 'other' to the Western Legal Tradition (WLT), such as mainland China.

Human Dignity in China: Etymological Roots, Philosophical Traditions and Current Interpretations

The reference to 'human dignity' as a Western concept should come as no surprise. In fact, the term *renge zunyan* (人格尊严)—which first appeared in Chinese statute law with the enactment of Article 38 of the 1982 Constitution, and is now included in Articles 109 and 990 of the Civil Code—is a neologism that was only introduced into the Chinese language at the beginning of the twentieth century (Wei 曲 1991). It should be noted, however, that the novelty pertains mainly to the term that acts as a determinant in it (renge 人格) (Zhai 2022), whilst the noun determined (zunyan 尊严) has a very long history. In particular, the origins of character zun 尊can be traced back to the Shang dynasty, when it was used to refer to the wine vessel employed by the sovereign for religious sacrifices. Over time, this character took on the meaning of 'noble' in the dual sense of the elevated status of a class of natural or human beings, and the attitude to be adopted towards this class—i.e., respect (Chia-Shin Hsu 2022). The character yan 严 also conveys the idea of respect and refers to solemn and dignified behaviour. Its combination with zun to signify noble status and dignified behaviour deserving of respect can be found in classical literature, beginning with the works of Xunzi (316–237 BC), one of the earliest Confucian philosophers. This two-syllable compound soon gained a central position in both Confucian ethical reflection and the two traditional indigenous Chinese doctrines: Daoism and Monism. In recent years, it is precisely the concept of dignity espoused by these philosophies that has been analysed and studied, to identify the cryptotypes underlying the interpretation given to the Chinese contemporary legal system.

Chinese scholars remark that, according to Confucian thought, dignity manifests itself primarily within a social context through an individual's mode of relating to others or to the group as a whole (Luo 2014). Confucianism recognises two types of dignity: innate dignity, which belongs to all individuals, and acquired dignity, which represents the virtue that each individual achieves in developing their moral potential (Li 2022a). Linked to these concepts is the notion of 'integrity' of human dignity, interpreted as the need to preserve the capacity to develop virtues. This capacity exists thanks to an awareness of the finite and limited nature of humanity (Jordan 2010). From this perspective, therefore, biotechnologies that allow these limits to be overcome would be ethically reprehensible as they contradict 'dignity', in the dual meaning attributed to it by Confucianism (Li 2022a). In Daoism, on the other hand, the relevant concepts from the perspective of biotechnology are abstinence and non-intervention (*wuwei* 无为). This implies abandoning worldly desires, competition, and the thirst for knowledge and ethical precepts. The aim is to reduce one's needs to the essence, so as not to hinder the perpetual flow of the universe and, consequently, its dignity (Qiao 2014). As regards Moist thought, the greatest contribution to current bioethical discourse could be provided by the link between the notion of dignity and the concept of 'undifferentiated love', often translated as 'universal love' (*jian'a*i 兼爱) (Zhang 2016). In fact, undifferentiated love not only recognises a minimum amount of human dignity for all human beings, regardless of their circumstances or gender, but is also realised in concrete terms through the achievement of profit (*li* 利). This profit is understood as that which benefits the whole world and is measured by three parameters, always referring to the community: material prosperity, a population or family of acceptable size, and personal and family security (Fraser 2016). According to some authors, it is exactly in providing criteria based on collective advantage and measured in terms of concrete benefits for the greatest number of people that this Moist notion of dignity could prove useful today, inspiring the regulation of life-related technologies (Xiao萧 2018).

It is worth mentioning that Chinese legal scholars and political leaders usually do not explicitly refer to the Chinese philosophical tradition in current discussions about human dignity. On the contrary, they tend to emphasise the foreign origin of the concept, citing German jurisprudence, case law and even Article 1 of the Basic Law of the Federal Republic of Germany (Novaretti 2024). However, they cannot help but express the concept from a perspective that focuses on the relationships between individuals rather than on individuals themselves (Novaretti 2024). To me, this constitutes more than a minor nuance, particularly in the context of controlling the risks associated with biotechnology. By its very nature, biotechnology is border-insensitive, both in terms of states and with regard to individuals. Therefore, an approach that views human beings as part of a network of relationships and interprets dignity as a limit and a parameter against which to assess whether a given scientific development can deliver concrete benefits to the greatest number of people is certainly noteworthy.

Conclusions

This chapter analyses the origins of prejudices regarding the irrelevance of law in mainland China, examining their evolution from the perspective of the country's legal history, from the first encounter between European and Chinese legal systems in the nineteenth century, up to the enactment of the PRC's Civil Code in 2020. Particular reference is made to the 'Western' legal models adopted by Chinese reformers in various historical periods, remarking how efforts to modify its legal system have failed to change the perception of China as a country without law, except superficially. Not even the enactment of the Civil Code has changed this view significantly. This is despite the fact that the Code testifies to the maturity and sophistication achieved by the Chinese legal system, adopting original solutions with regard to structure, language, fundamental institutions, and related matters, whilst maintaining the German code as the main reference from a legal comparative point of view. In particular, the Civil Code contains rules referring to both Chinese traditional and socialist values and—following the example of the most recently enacted Civil Codes—contains provisions for the defence of fundamental rights, dedicating an entire book to personality rights, which it protects in both real and virtual life. For this reason, I do not consider Chinese scholars to be entirely wrong when they describe it as 'a twenty-first-century code for a new era, based on Chinese national conditions and characteristics'. However, can the combination of foreign and Chinese characteristics reflected in the Code also contribute to global risk governance?

In my view, a possible starting point for a Chinese contribution to biogovernance commons can be the principle of human dignity included in the Arts. 109 and 990 of the PRC Civil Code. In fact, by including reference to this notion (of 'Western' origin) in the highest source of civil norms, the Chinese legislator has made it easier for PRC judges to use it as a basis for decisions regarding not yet specifically regulated biotechnologies, as is already happening in cases relating to the clinical applications of biotechnologies related to the beginning of human life, such as post-humous fecundation and gestational surrogacy (Novaretti 2025). It seems to me that the Chinese understanding of the concept, which considers individuals in a relational way, as part of social groups which encompass family, society, the state, and humanity as a whole, can be usefully taken into consideration at the global level to build a biogovernance system which includes and mediates between cultural and legal traditions that differ from the Western one.

There is, however, a need for an open, transparent, and unprejudiced discussion on these issues, both inside and outside China. In particular, without knowledge, there can be no understanding. For this reason, I believe that China must also take certain steps, starting with small things. Reflecting on my daily experience as a scholar of Chinese law based outside China, I cannot help but notice that researching China has become increasingly complicated over the years. This refers to the ease with which information, news, and texts of judgments issued by the People's Courts disappear from the web overnight; the impossibility of finding regulations that have been in force for more than 20 years in regularly subscribed Chinese legal databases; and the difficulty of accessing Chinese scientific publications even when they appear to be available online. These may be seen as minor, harmless details. However, the fact that they concern not research on politically sensitive issues or state secrets, but rather 'basic' information on data relating to the legal theory and daily practice of one of the world's most important countries from geopolitical, economic, and scientific viewpoints, shows that prejudice and mistrust are not only Western characteristics.

If we truly want to create global governance for the common good, in and beyond science, we must limit barriers, not erect them where they are not needed. Without a community of shared information, ideas, and solutions, there cannot be a shared future for mankind.

Bibliography

Albers, Marion. 2018. Biotechnologies and human dignity. In *Human dignity in context: Explorations of a contested concept*, edited by Christoph Möllers, Dieter Grimm, and Alexandra Kemmerer. Oxford: Baden-Baden.

Andorno, R. 2009. La tutela della dignità umana: fondamento e scopo della Convenzione di Oviedo. In *Zurich Open Repository and Archive*. https://doi.org/10.5167/UZH-18746. Accessed 18 Feb 2026.

Baskir, Cecily E. 2015. Legal education in China: Globalizing with Chinese characteristics. *Asian Journal of Legal Education* 2 (2): 143–156. https://doi.org/10.1177/2322005815578516.

Brownsword, Roger. 2014. Human dignity from a legal perspective. In *The Cambridge handbook of human dignity*, edited by Marcus Düwell, Jens Braarvig, Roger Brownsword, and Dietmar Mieth, 1a ed., 1–22. Cambridge University Press. https://doi.org/10.1017/CBO9780511979033.003.

Chen, Jianfu. 2015. *Chinese law: Context and transformation*. Expanded, Revised ed. Leiden: Martinus Nijhoff.

Chen, Li. 2024. Concretising the legal professional community in late imperial China, c. 1700–1900. *Asian Journal of Law and Society* 11 (3): 260–287. https://doi.org/10.1017/als.2024.20.

Chen, Zhiwu, and Kaixiang Peng. 2022. Production, consumption, and living standards. In *The Cambridge economic history of China: Volume 1: To 1800*, The Cambridge economic history of China, edited by Debin Ma and Richard von Glahn, vol. 1. Cambridge: Cambridge University Press. https://doi.org/10.1017/9781108587334.020.

Chia-Shin Hsu, Jimmy. 2022. Human dignity in the jurisprudence of the Taiwan constitutional court. In *Human dignity in Asia: Dialogue between law and culture*, edited by Jimmy Chia-Shin Hsu. Cambridge: Cambridge University Press. https://doi.org/10.1017/9781108886598.006.

Clarke, Donald C. 2020. Anti anti-orientalism or is Chinese law different? *American Journal of Comparative Law* 68 (1): 55–136. https://doi.org/10.1093/ajcl/avaa004.

Derrida, Jacques. 2016. *Of grammatology*. 4th ed. Baltimore: John Hopkins University Press.

Fraser, Chris. 2016. *The philosophy of the Mòzǐ: The first consequentialists*. New York: Columbia University Press.

Granet, Marcel. 1934. La pensée chinoise. Paris: Albin Michel.

Hao, Jiang. 2021. The making of a civil code in China: Promises and perils of a new civil law. *Tulane Law Review* 95 (4): 777–820.

Hao, Jiang, and Pietro Sirena. 2023. *The making of the Chinese civil code: Promises and persistent problems*. Cambridge: Cambridge University Press. https://doi.org/10.1017/9781009336611.

Head, John W., and Yanping Wang. 2005. *Law codes in dynastic China. A synopsis of Chinese legal history in the third centuries from Zhou to Qing*. Durham: Carolina Academic Press.
Hegel, Georg Wilhelm Friedrich. 2001. *The philosophy of history*. Kitchener: Batoche Books.
Huang, Philip C. C. 2001. *Code, custom, and legal practice in China: The Qing and the republic compared*. Redwood City: Stanford University Press.
Information Office of the State Council. 2011. The socialis system of laws with Chinese characteristics. State Council of The People's Republic of China. http://english.www.gov.cn/archive/white_paper/2014/09/09/content_281474986284659.htm. Accessed 4 Nov 2025.
Jones, William C., Tianquan Cheng, and Yongling Jiang. 1994. *The great Qing code: A new translation*. Oxford: Oxford University Press.
Jordan, Matthew C. 2010. Bioethics and 'human dignity'. *The Journal of Medicine and Philosophy* 35 (2): 180–196. https://doi.org/10.1093/jmp/jhq010.
Li, Yaming. 2022a. The confucian concept of human dignity and its implication for bioethics. *Developing World Bioethics* 22 (1): 23–33. https://doi.org/10.1111/dewb.12312.
Li, Zaiquan. 2022b. The reform of the legal system in the late Qing. *Journal of Modern Chinese History* 16 (1): 50–70. https://doi.org/10.1080/17535654.2022.2101792.
Luo, An'xian. 2014. Human dignity in traditional Chinese confucianism. In *The Cambridge handbook of human dignity: Interdisciplinary perspectives*, edited by Dietmar Mieth, Jens Braarvig, Marcus Düwell, and Roger Brownsword. Cambridge University Press. https://doi.org/10.1017/CBO9780511979033.021.
Mallapaty, Smriti. 2018. Engineering a biomedical revolution. Nature 564(7735), S66–68. https://doi.org/10.1038/d41586-018-07692-4
Marx, Karl. 1858. Trade or opium?. *New York Daily Tribune* (New York), September 20. https://moodle2.units.it/pluginfile.php/101065/mod_resource/content/2/Marx_Writings_on_China_NYDT_1853-60.pdf. Accessed 4 Nov 2025.
Novaretti, Simona. 2019. China's sustainability challenges: Confucianisation of Chinese law from intra-generational, inter-generational and gender equity perspectives. *NAVEIÑ REET: Nordic Journal of Law and Social Research* 9 (December): 21–60. https://doi.org/10.7146/nnjlsr.v1i9.122149.
Novaretti, Simona. 2024. Dignità umana e nuove tecnologie. La regolamentazione dell'ingegneria genetica nella Repubblica Popolare Cinese, tra valori tradizionali, modelli giuridici occidentali e nuovi orizzonti delle bioscienze (Human dignity and new technologies. The regulation of genetic engineering in the people's republic of China, between traditional values, Western legal models, and new biosciences' horizons). *Daimon—Diritto Comparato delle Religioni, Quaderni di Diritto e Politica Ecclesiastica* I:191–216. https://doi.org/10.1440/114451.
Novaretti, Simona. 2025. *Law and bioethics: The beginning of life in China*. Leiden: Brill.
Qiao, Qingju. 2014. Dignity in traditional Chinese daoism. In *The Cambridge handbook of human dignity: Interdisciplinary perspectives*, edited by Dietmar Mieth, Jens Braarvig, Marcus Düwell, and Roger Brownsword. Cambridge University Press. https://doi.org/10.1017/CBO9780511979033.022.
Roy, Olivier. 1972. *Leibniz et la Chine*. Paris: Librairie Philosophique J Vrin.
Ruskola, Teemu. 2002. Legal orientalism. *Michigan Law Review* 101 (1): 179–234.
Ruskola, Teemu. 2013. *Legal orientalism. China, the United States, and modern law*. Cambridge: Harvard University Press.
Tatlow, Didi Kirsten. 2015. A scientific ethical divide between China and West. *The New York Times*, June 29. https://www.nytimes.com/2015/06/30/science/a-scientific-ethical-divide-between-china-and-west.html. Accessed 4 Nov 2025.
van Beers, Britta C. 2020. Rewriting the human genome, rewriting human rights law? Human rights, human dignity, and human germline modification in the CRISPR era. *Journal of Law and the Biosciences* 7 (1): lsaa006. https://doi.org/10.1093/jlb/lsaa006.
Wang, Zhiqiang. 2024. Law and political power structure: Justice in early imperial China and Rome. *Yale Journal of Law & the Humanities* 35 (3): 626–642. https://openyls.law.yale.edu/entities/publication/bdacb98b-a6e8-43a3-a293-f133c2383d89.

Weber, Max. 1968. *The religion of China: Confucianism and Taoism*. Los Angeles: The Free Press.
Wei 曲, Qu 炜. 1991. *Renge Zhi Mi 人格之谜 (The Enigma of Personality)*. Beijing 北京:Zhongguo renmin daxue chubanshe 中国人民大学出版社.
Woolston, Chris. 2023. Nature Index annual tables 2023: China tops natural-science table. Nature, June. https://doi.org/10.1038/d41586-023-01868-3
Xiao萧, Hong'en 宏恩. 2018. You Moxue duanshi dangdai yiliao kejixia renzhi shenming zunyan 由墨学端视当代医疗科技下人之生命尊严 (the dignity of human life in contemporary medical technology from the perspective of Mohism). *Wuhan keji daxue xuebao* 武汉科技大学学报 20 (6): 635–640.
Zhai, Xiaobo. 2022. Personal dignity under Chinese: Constitutional law. In *Human dignity in Asia: Dialogue between law and culture*, edited by Jimmy Chia-Shin Hsu. Cambridge: Cambridge University Press. https://doi.org/10.1017/9781108886598.011.
Zhang, Qianfan. 2016. Equality and universal love: Human dignity in Mohism. In *Human dignity in classical Chinese philosophy: Confucianism, Mohism, and Daoism*. New York: Springer. https://doi.org/10.1057/978-1-349-70920-5_5.
Zhou周, Youjun友军. 2020. Minfa Dian Renge Quan Bian Pinxi 民法典人格权编评析 (Compilation and Analysis of Personality Rights in the Civil Code). Zhongguo Minshang Falü Wang 中国民商法律网 (China Civil and Commercial Law Net). https://www.civillaw.com.cn/t/?id=36942. Accessed 4 Nov 2025.

Simona Novaretti is an Associate Professor of Comparative Private Law at the University of Turin, School of Law, where she teaches Chinese Law, Comparative Law, and Law and Society in Asia. She graduated and obtained her Bachelor and Master's Degree in Chinese Language and Literature from the University of Venice, Ca'Foscari, got her LLB and LLM from the University of Turin, School of Law, and got her Ph.D. in Comparative Law from the University of Milan. Her research activity focuses on Chinese Law. She has published articles on Chinese law and language, Chinese contract law, Chinese public interest law and Chinese law and society. Her latest monography explores the relationship between law and bioethics in China with a particular focus on the regulation of biotechnologies related to the beginning of human life (Brill 2025). She has published monographs on public interest litigation in China and on public participation in the protection of cultural heritage in the PRC. Simona is member of the Bioethical Committee of the University of Turin. She is member of the scientific committee of the Institute of Asian Studies (ISA), of the Interdisciplinary Center for Research and Studies of Women and Gender (CIRSDe) and of the Interdipartimental Center Labont—Laboratory for Ontology. She is also member of the European Chinese Law Association; of the Italian Association of Comparative Law; of the Italian Society for the Research on Comparative Law; of the Italian Association of Chinese Studies; and of the European Law Institute.

Chapter 7
Rethinking Global Biosecurity Governance: Emerging Non-Western Perspectives

Suryesh Kumar Namdeo

Abstract This chapter maps the complex and evolving landscape of biosecurity governance, tracing major actors, key challenges and disproportional Western influence. It discusses how effective governance requires coordination among diverse stakeholders and across multiple levels, from laboratories and government agencies to regional networks and multilateral institutions. The chapter analyses the challenges of existing governance systems, including inequitable representation, uneven capacity, siloed operations and Western dominance in setting norms and agendas. To address these challenges, it highlights BioGovernance Commons as an emerging and inclusive framework based on sustained dialogue, relational norms and mutual accountability. Further, the chapter elaborates diverse non-Western biosecurity governance practices through four case studies involving China, India, Nigeria and Singapore and provides several recommendations for effective biosecurity governance. It underscores the importance of understanding engagement with diverse perspectives through the BioGovernance Commons framework to achieve more equitable, context-sensitive and collaborative biosecurity governance.

Keywords Biosecurity · Biorisks · BioGovernance commons · Biosafety

Introduction

The last decades have witnessed enormous progress in life sciences, which has contributed significantly to health, agriculture and the economy. The barriers to knowledge, expertise and resources to conduct biological research have come down in the process. However, several potential risks have emerged alongside the enormous benefits of these developments. Here, the risks include the potential deliberate

S. K. Namdeo (✉)
Indian Institute of Science, Bengaluru, India
e-mail: suryeshn@iisc.ac.in

J. Y. Zhang et al. (eds.), *BioGovernance Commons and China's Role in Global Risk Governance*, Emerging Globalities and Civilizational Perspectives,
https://doi.org/10.1007/978-3-032-32198-5_7

biological threats, including the development, production and use of biological weapons. Preventing such deliberate biological threats requires effective biosecurity governance. Biosecurity governance has been described as the institutional processes, policies and practices that shape behaviour to prevent or deter the misuse of biological science and technology (Evans et al. 2020). Biosecurity governance covers relevant laws, regulations, norms, best practices, institutional structures and coordination mechanisms to ensure that research and innovation in life sciences are done in a safe and secure manner. Biosecurity governance is crucial as it provides rules, practices and institutions to manage biological risks responsibly and collectively, while ensuring the growth of the bioeconomy (Trump et al. 2021). So, an effective biosecurity governance framework should have provisions for correct classification of biological incidents, assignment of authority and responsibility, investigation and prosecution, regulation of bioagents and international compliance and cooperation. It would require an integrated approach involving diverse stakeholders throughout the process of prevention, detection and mitigation of biosecurity risks (NTI 2025; Rawluk et al. 2021).

Here, BioGovernance Commons is emerging as a key conceptual framework under which biosecurity governance could be conceived as a continuous practice through *commoning* that involves acts of support, conflict negotiation and experimentation that shape transnational knowledge systems through ongoing relations among individuals (Zhang et al. 2022). This approach focuses on sustained dialogue involving key relevant stakeholders from diverse backgrounds to share perspectives on the shared challenges. This approach also foregrounds differences such as culture, capacity and values for building trust and mutual understanding through ongoing exchanges. The key advantage of this approach is that it focuses more on the ongoing multistakeholder engagement, ensures diverse representation and relational norms as outcomes instead of one-size rules. All of these become crucial for effective biosecurity governance as described in the passages below.

This chapter elaborates on the levels of biosecurity governance, major stakeholders involved, challenges in implementation, dominance of Western narratives and emerging non-Western perspectives. Further, the biosecurity governance components analysed in this chapter include relevant national laws, rules and regulations, national policy or strategy pertaining to deliberate and accidental biothreats, implementation agencies and mechanisms, self-regulatory approaches and programmes for international cooperation. Together, they provide a comprehensive picture of the current state of affairs with biosecurity governance across the world and highlight the importance of the BioGovernance Commons approach for ensuring mutual trust, balanced representations and relational norms.

Multi-Stakeholder Engagement for Biosecurity Governance

There are different stakeholder groups that need to be engaged for effective biosecurity governance (Reed and Curzon 2015). These include government officials across different levels and ministries; academics including researchers and research institutions; civil society actors such as think tanks and NGOs; industry actors such as biotech companies and industry associations; and international institutions and frameworks like the Biological Weapons Convention (BWC), the World Health Organisation (WHO) and the Global Health Security Agenda (GHSA). Each stakeholder group has their own strengths, limitations and interests, which are often complementary (Namdeo 2026). For example, the government stakeholders' strength lies in their authority, resource availability and crisis response capabilities, while their limitations often include weak coordination among different agencies and limited adaptability to rapid technological changes. Their major interest lies in ensuring public safety and maintaining law and order during a crisis.

Similarly, industry stakeholders' strength lies in their ability to innovate and create new products, including potential surveillance tools and countermeasures against biothreats, while their limitation here is their profit-driven nature, which is generally averse to regulations that could potentially reduce biothreats. Their major interest is in accelerating innovations to improve pandemic preparedness, while reducing regulations and ensuring strong financial gains. It is important to analyse the respective strengths, limitations and interests of different stakeholders to develop effective biosecurity governance solutions. Since each stakeholder has unique strengths and limitations, it is necessary to promote collaborative efforts, as no single stakeholder can fully address biosecurity governance challenges. Here, the BioGovernance Common approach could help establish collaborations by building trust and breaking siloes among stakeholders through regular dialogues. However, such an effort would require engaging stakeholders across different levels of biosecurity governance.

Levels of Biosecurity Governance

An effective biosecurity governance would require concrete strategies and efforts at different levels to efficiently and flexibly minimise biosecurity threats. Here, the first level is research labs and institutes. Here, several mechanisms with overlapping roles, like the institutional biosafety committees, institutional review committees or institutional bioethics committees, exist around the world. These committees are responsible for ensuring biosafety, biosecurity and bioethics at research institutes and consist of internal and external experts and biosafety professionals, who conduct regular reviews and provide approval for certain projects and experiments. In addition, research institutions in most countries have biosafety officers and/or regulatory compliance officers supported by an assigned person in each lab to ensure that the biosafety practices are followed. The effectiveness of biosecurity governance at this level depends on various factors, including the composition of the committees and the inclusion of the right expertise, the degree of understanding of the evolving biosecurity challenges among biosafety officers and the right use of reporting mechanisms for biosafety/biosecurity vulnerabilities and incidents.

The second and probably the strongest level of biosecurity governance is at the level of provincial and national governments. Here, government authorities create regulations and oversight mechanisms like the US Federal Select Agent Program and the Chinese Biosecurity Law to effectively govern biosecurity in their jurisdictions (Federal Select Agent Program 2025, China Daily 2021). In most cases, countries lack a unified structure or mechanism to address diverse biosecurity challenges and governance efforts are fragmented across several government agencies. Government agencies also regulate the biosecurity landscape through their funding decisions and requirements posed by funding agencies. Further, due to their authority, they possess strong convening power to get other stakeholders together for implementing biosecurity measures.

The third level for biosecurity governance is regional, involving regional governance structures like the European Union, ASEAN and the African Union. Here, it is important to note that not all regions have coherent and functional governance structures, and several regions have not yet prioritised biosecurity. However, there are some successful examples of biosecurity governance at this level. For instance, the Biosafety and Biosecurity Initiative of Africa Centres for Disease Control and Prevention (Africa CDC) supports member states to strengthen biosecurity systems in the region, help them fulfil their multilateral obligations and help develop National Action Plans for Health Security (NAPHS) (Africa CDC 2025b). The European Union (EU) has created extensive regulations on biosafety and food safety and is currently deliberating on incorporating biosecurity in the upcoming EU Biotech Act (European Biosafety Association 2026, IBBIS 2025). In Southeast Asia, ASEAN leaders issued a joint statement in 2024 on strengthening regional biosafety and biosecurity (ASEAN 2024). As next steps, ASEAN member states are currently finalising a plan of action for the ASEAN Network on Biosafety and Biosecurity (ASEAN Hub 2025).

The fourth level is multilateral, which involves governance at the level of the United Nations and other multilateral bodies such as G20 and BRICS. The existing major multilateral frameworks and instruments for biosecurity are the Biological Weapons Convention, the United Nations Security Council Resolution 1540, the International Health Regulations (IHR), the Australia Group and the United Nations Secretary-General's Mechanism for Investigation of Alleged Use of Chemical and Biological Weapons (UNSGM). Each of them has certain strengths and limitations, and existing complementarities to potentially create a web of prevention (Vennis et al. 2022, Novossiolova et al. 2021). However, their effectiveness vastly depends on States to enact and implement biosecurity legislation and regulations domestically. Globally, the resources allocated to biological disarmament and biosecurity are several orders of magnitude less than their nuclear and chemical counterparts (Revill 2024). Non-UN multilateral settings like G20, BRICS, the Shanghai

Cooperation Organisation (SCO), and Global Health Security Agenda (GHSA) provide other potential opportunities for biosecurity governance that are not yet fully realised.

In addition to these levels, there are several existing community statements, self-regulations, guidelines, norms, and codes of conduct that support biosecurity governance. For example, the International Gene Synthesis Consortium (IGSC) is an industry-led effort to reduce biosecurity threats due to synthetic nucleic acids by developing and applying screening protocols as a self-regulatory approach (IGSC 2026). In another effort, a community statement titled 'Community Values, Guiding Principles, and Commitments for the Responsible Development of AI for Protein Design' is released by scientists, biosecurity experts and AI developers to reduce AI-Biosecurity risks (Responsible AI X Biodesign Community 2024). Similarly, AIxBio Global Forum is convened as a multi-stakeholder platform to deliberate on emerging risks and potential policy solutions at the AI-biosecurity interface (NTI 2024a). From the academic side, the Tianjin Biosecurity Guidelines for Codes of Conduct of Scientists, championed by China, aims to create norms for the scientific community for strengthening biosecurity (Wang et al. 2021). Lastly, the UN 'Youth Declaration for Biosecurity' calls for strengthening biosecurity by empowering young professionals and supporting their involvement in biosecurity roles. Here, the process of *commoning* through the development of relational norms could play a crucial role in reducing tensions among different initiatives in building mutual understanding.

Major Challenges in Biosecurity Governance

There are several issues negatively affecting effective biosecurity governance. These include the presence of a siloed approach across different governance levels, agencies and stakeholders. This is further complicated by the lack of understanding of each other's strengths, limitations and interests, which results in sub-optimal collaborations. Currently, funding is extremely limited and uneven for both policy and scientific research to develop solutions to technical and policy challenges. The funding priorities vary widely among countries and funding agencies. In general, the focus in the West has been distributed between prevention and detection against biosecurity risks, whereas the focus is more on the response side in several other parts of the world.

There is also a limited awareness about deliberate and accidental biothreats and their potential impact among all stakeholder groups, especially in low- and middle-income countries. This affects their ability to understand the evolving threat landscape and develop potential responses. Further, the existing multilateral and international frameworks and conventions to reduce biological threats, such as the BWC and IHR, lack strong compliance mechanisms. This is also true for the national regulations, as it is often difficult to distinguish peaceful research from dual-use research, complicating regulatory compliance. Further, due to underinvestment in research and innovation, there is limited capacity to analyse threats and develop countermeasures. Similarly, there are very few training and capacity-building programs to provide adequate training in various aspects of biosecurity risk management and mitigation. Lastly, there are a few standards and certification programs at the national and international levels to ensure safety and security in a harmonious, globally recognised manner.

Dominance of Western Perspectives in Biosecurity Governance

The West, particularly the USA and the UK, dominate the biosecurity governance discourse as they have historically set the agenda, guidelines, norms and standards in biosecurity. These include leading deliberations to establish the Geneva Convention (1925) and later the Biological Weapons Convention. In recent times, the focus has shifted to lead the establishment of norms and standards to govern nucleic acid synthesis, AI-biology interface and Mirror Life research. Notably, the USA has nearly half of the BSL-3 labs in the world, the UK has the highest number of BSL-3 labs per million people, and Europe has 46% of BSL-4 labs in the world (Muluneh et al. 2025). The USA and the UK also account for the majority of the academic publications in biosecurity, with several European countries counting among other top producers of biosecurity literature (Namdeo and Chandan 2025). Also, security frameworks like the G7-led Global Partnership Against Weapons of Mass Destruction, and the health initiatives such as the US-led Global Health Security Agenda (GHSA) and the Coalition for Epidemic Preparedness Innovations (CEPI) fund, guide and shape major capacity-building activities in the resource-constrained settings worldwide.

The USA and the UK also have influential domestic biosecurity policies and programs that shaped broader international narratives and funding. The USA, for example, has a well-resourced, multi-layered biosecurity and biodefense architecture involving numerous federal agencies, professional bodies, academic centres and civil society organisations (The White House 2022; Bipartisan Commission on Biodefense 2024). It also has a robust legal, policy and regulatory apparatus to address deliberate biological threats (The White House 2022, Congressional Research Service 2024). The USA has several long-standing federal programmes that have had global influence and shaped global norms and best practices. These include the Federal Select Agent Program, Federal policy for Oversight of Dual Use Research of Concern, National and Regional Biocontainment Labs, the Biosecurity Engagement Program of the Department of State and the Biological Threat Reduction Program of the Department of War (Federal Select Agent Program 2025, U.S. Department of Health and Human Services 2024, NIAID 2026, US Department of State 2026, National Academies of Science Engineering and Medicine 2020). Similarly, the UK Biological Security Strategy and its implementation have helped create norms to prepare multi-faceted national biosecurity strategies and implementation roadmaps (UK Government 2025).

The Western biosecurity policies and strategic documents focus strongly on surveillance, prevention, oversight, deterrence, attribution and global leadership (Green and Ananthasekar 2023; Bielecka-Oder 2018; Linkous et al. 2021). They call for implementing well-funded, long-term biosecurity and biodefense programmes, which go beyond the domestic requirements and focus on global norms and standard setting to reduce global biological threats as prioritised by the West (Morrison and Simoneau 2023; Justen et al. 2023). These strategies are further strengthened by the numerous Western research institutions, think tanks and advocacy bodies. Their influence could be witnessed clearly in the NGO statements and side events organised during the official BWC meetings. These efforts are complemented by biosecurity projects and policy advocacy funded and conducted around the world by philanthropic funding bodies and NGOs that are mostly headquartered in the North Trans-Atlantic (C.K. 2024). This 'West to rest' pattern for biosecurity governance, with a few developed countries in the West and Global North defining norms and standards for the rest of the world, creates inherent inequities that need to be balanced with more diverse global efforts (Namdeo and Zhang 2024; Prouse and Arefin 2025).

Importance of Non-Western Perspectives

It is often emphasised that there are distinct biosecurity challenges in low-resource settings, and these can bring crucial perspectives to develop inclusive and equitable global policy (NTI 2024b). Involving such countries in making and shaping the rules, norms and standards is crucial to tailor the safeguards to real-world conditions and build broader buy-in. Further, a diverse and inclusive representation in international biosecurity governance is crucial for avoiding colonial biases and for incorporating practices from different knowledge systems, including indigenous practices and local bio-surveillance networks. It is also important to remember that while there are many overlaps in biothreat profiles in regions around the world, there are also significant differences. For example, Yellow fever is a menace in many parts of Africa, but most Western countries are unaffected by it and don't prioritise it. In general, Africa, the Caribbean, South Asia and Southeast Asia have frequent zoonotic spillovers of diseases like Nipah virus, Lassa fever and avian influenza, and preventing and mitigating them is a major priority in these regions. In contrast, risks at the AI-biosecurity interface and Mirror Life research have emerged as major concerns among developed countries and have received significant funding. However, these remain low priorities for most developing countries as they have immediate concerns like providing robust safeguards to their BSL3 and BSL4 facilities and strengthening their institutional biosafety/biosecurity committees.

The biosecurity challenges that several developing non-Western countries face are the lack of sufficient high containment biosafety facilities, limited lab audits, absence of lab and personnel certification programmes, inadequate coordination among government agencies, limited biosurveillance capabilities, insufficient stockpiles of medical and non-medical countermeasures, lack of adequate funding for biosecurity and major gaps in policy and regulatory frameworks. Several strategies, practices and innovations are taking shape in these regions and countries to address these challenges. Analysing, contextualising and understanding these solutions would be crucial for effective global biosecurity governance. Here, the BioGovernance Commons approach could provide an established framework through which regular multi-stakeholder dialogues could be held between Western and non-Western countries. Overall, diverse non-Western perspectives will enrich the current biosecurity discourse by bringing up the challenges and solutions often overlooked by the existing lopsided, West-dominated approaches.

Case Studies on Emerging Non-Western Perspectives

China

The People's Republic of China has a very comprehensive national biosecurity law (2021) that highlights biological threats from a national security perspective and divides them into two broad categories: naturally occurring risks and new risks brought about by the development of biotechnology, which include both accidental and deliberate biothreats (Liang et al. 2021). It establishes several elaborate systems and coordination mechanisms for the prevention and response of biothreats (Xiao et al. 2023). Also, Chinese Criminal Law and Export Control regulations contain clear provisions for regulating dual-use biological agents, toxins and related equipment (Min et al. 2025). Further, the Regulations on the Safety Management of Biotechnology Research and Development (draft) are considered among the critical pieces developed in conjunction with the promulgation of the Biosecurity Law and are intended to serve as a reference for individuals involved in biotechnology R&D and application (Ministry of Science and Technology of the People's Republic of China 2019). Overall, the Chinese approach is State-centric, top-down coordination via the State Council and the National Health Commission. It is also precautionary with a focus on risk prevention, ethical oversight and national control, while having a broad scope across health, agriculture, ecology, national security and biotechnology.

China engaged in in-depth discussions with various countries and reached the Tianjin Biosecurity Guidelines for Codes of Conduct for Scientists in July 2021, putting forward ten major guiding principles, including ethical standards, laws and norms, responsible conduct of research, research process management and international cooperation (Qiushi 2023). It also actively supports the publication of local institutions' biosecurity research outcomes in the Journal of Biosafety and Biosecurity, published by the Chinese Academy of Sciences (Lee 2023). China is an active member of the BWC and contributes to it through various working papers, expert consultations and diplomatic support in other forums such as the Shanghai

Cooperation Organisation (Ministry of Foreign Affairs of the People's Republic of China 2024). In general, China is increasingly active in global health and biosecurity diplomacy and has emerged as a leading voice in shaping biosecurity governance on several key issues (UNODA 2026, Dale-huang and Beauchamp-Mustafaga 2024).

India

India has a large and rapidly growing bioeconomy with several government programmes and institutions to support it (BIRAC 2024). India currently lacks a specific law or strategy for biosecurity, but its legislative framework, including the Environment Protection Act (1986), the Weapons of Mass Destruction Act (2005) and the Disaster Management Act (2005), could partially address biosecurity challenges. India also has integrated regulatory mechanisms, such as the Review Committee on Genetic Manipulation (RCGM) and Institutional Biosafety Committees (IBSCs), that oversee approval of all genetic modification experiments (Department of Biotechnology, Ministry of Science and Technology n.d.). Further, the Indian Biosafety Knowledge Portal (IBKP) contains information and regulatory guidance regarding biosafety, biocontainment and genetically modified organisms (GMOs) (DBT, Government of India 2026). It also records information collected regarding RCGM and IBSCs. The IBSCs, in particular, conduct risk assessment and are empowered to make decisions regarding biosafety and biosecurity aspects of any genetic experiments to be conducted in their respective institutes. Also, all institutes and research labs conducting genetic experiments must have IBSCs, and their decisions and reports must be shared with the Central government. Additionally, the National One Health Mission provides a coordination mechanism across several government agencies for pandemic preparedness and prevention, integrated disease surveillance, outbreak investigation and a data system that could contribute significantly to prevention, detection and response of biosecurity threats (Office of the Principal Scientific Adviser to the Government of India 2026).

India has international obligations managed by the national contact points for several biosecurity-related multilateral instruments such as the BWC, the Geneva Protocol, the International Health Regulations, GHSA, UNSCR 1540 and the Australia Group. India's export control system has extensive regulations for the 'Special Chemicals, Organisms, Materials, Equipment and Technology' (SCOMET) list in accordance with its international obligations (Directorate General of Foreign Trade 2018). In December 2025, India organised a 2-day international conference on 50 Years of BWC: Strengthening Biosecurity for the Global South' that involved participation from over 8o countries along with diverse Indian stakeholders (Ministry of External Affairs, Government of India 2025). Some new capacity-building initiatives like the Bio-responsibility Future Leaders Programme (BFLP) have started to build national capacity against biological threats (IndiaBioscience

2026). Overall, the Indian approach is a combination of bottom-up and top-down mechanisms with IBSCs and the National One Health Mission playing crucial roles.

Nigeria

The Federal Republic of Nigeria is a leading major Global South country that has prepared a comprehensive National Biosecurity Policy and Action Plan. This policy has an explicit focus on One Health, combining biosecurity aspects related to human health, environment and agriculture and food safety in a holistic manner, while identifying key stakeholders engaged in different areas (NBMA 2022). It identifies the National Biosafety Management Agency (NBMA) as the focal point for coordination across different agencies and stakeholders to ensure biosecurity prevention and preparedness measures. It also mandates the establishment of a National Advisory Council and a Technical Working Group for annual biosecurity evaluation, coordination and decision-making. The national policy adopts a risk-based screening and prioritisation approach and provides a categorised list of priority agents and toxins. Notably, the policy recognises surveillance, training, simulation exercises, hotspot mapping and a national rapid response programme as core actions for national capacity building. Lastly, the policy acknowledges major gaps in funding and capacity and calls for multi-source financing, transparent financial management and a monitoring and evaluation framework.

Overall, the National Biosafety Management Agency is empowered to implement all domestic regulations as well as ensure the implementation of relevant multilateral conventions related to biosafety and biosecurity. Nigeria actively participates in several regional and multilateral efforts to strengthen biosecurity. For example, it actively supports regional biosafety and biosecurity efforts by the Africa Centre for Disease Control and Prevention (Africa CDC), the African Union (AU) and the Economic Community of the West African States (ECOWAS) (Maruta et al. 2023, Africa CDC 2025a, 2025b). Further, Nigeria is a State Party to the BWC, actively participates in the Global Health Security Agenda (GHSA) and implements the WHO International Health Regulations (IHR). Lastly, several Nigerian alumni of the UN Youth for Biosecurity have created a biosafety and biosecurity handbook for students that has become a key global resource for communicating this topic (BWC Implementation Support Unit n.d.).

Singapore

Singapore, with its unique governance model, has an effective whole-of-government approach to address the whole range of biothreats. This approach exemplifies diverse government agencies responsible for public health, biosafety, law enforcement, intelligence and science coming together to address emerging bio-threats

(Nuclear Threat Initiative and Johns Hopkins Center for Health Security 2021, Tan 2025). Singapore treats man-made biological threats as part of the broader Chemical, Biological, Radiological, Nuclear and Explosive (CBRNE) risk spectrum and aligns its system for both deliberate and accidental events. Singapore's legal framework covers the whole spectrum of biological threats from biosafety in research to prevention of bioterrorism and control of dual-use research. The most important pillar of this legal framework is the Biological Agents and Toxins Act (BATA) 2005, which contains provisions for licensing and bio-facility approval, bio-facility registration, transfer and transport of biological material, disposal and decontamination, physical and personnel security measures, incident reporting and enforcement (Ministry of Health, Singapore 2005). Further, the Internal Security Act 1960 and the Terrorism (Suppression of Financing) Act 2002 empower the government to act against individuals and groups involved in activities such as bioterrorism or providing material support for such activities (Attorney-General's Chambers 1960, 2002). Singapore's Strategic Goods (Control) Act 2002 regulates export, transit and brokering of strategic and dual-use items, including biological material, equipment and technologies to prevent proliferation and misuse of potentially dangerous biological material (Singapore Customs 2026). Lastly, biosafety guidelines on GMOs complement BATA's focus on high-risk pathogens with controls on innovation risks.

Singapore has institutionalised biosecurity as part of the CBRNE national resilience strategy, anchored by the CBRNE Centre of Expertise (HTX 2026). It works with a specialised lab network to conduct research, develop sensors and simulate threats to keep Singapore ahead of the threat curve. Further, Singapore actively works towards strengthening ASEAN's collective capacity to prevent, detect and respond to biological and CBRN threats through the ASEAN Centre for Public Health Emergencies and Emerging Diseases and the Network of ASEAN CBR Defence Experts. Multilaterally, Singapore is a State Party to the BWC and an active member of the Global Health Security Agenda (GHSA). Finally, the Asia Centre for Health Security (ACHS) at the National University of Singapore has emerged as a key hub for biosecurity research and deliberation in the Asia-Pacific region (ACHS 2026).

Collectively, these case studies demonstrate successful non-Western biosecurity governance practices. The countries studied here are all working towards devising strategies and systems that are shaped by their own risk landscape and resources. The current approaches by these countries are independent of the BioGovernance Commons framework, but they, in spirit, share similar goals and show a range of options and practices developing. Further, recognising and popularising these strategies and systems through *commoning* could be key to enabling cross-learning and legitimising the non-Western contributions.

How to Strengthen and Promote Non-Western Perspectives in *Commoning*?

It is essential to have sustained, inclusive, multi-stakeholder efforts to promote Non-Western perspectives, given their importance as highlighted in the passages above. This would require, first, ensuring inclusive and balanced representation in expert groups and technical committees of multilateral bodies like the BWC, WHO and GHSA through quotas, and travel and knowledge access support. Second, the Non-western perspectives should be popularised through joint sponsorship of working papers and other internal documents during the official meetings of the BWC and other international bodies. Third, technical and policy research funding opportunities should be created that prioritise local leadership, co-design and capacity transfer. Here, co-funding possibilities with a local or national partner could be explored to better identify priorities and research proposals. Such funding opportunities could also integrate awareness-raising and capacity-building efforts along with research.

Fourth, equitable North-South and South-South research collaboration should be implemented through a science diplomacy framework involving joint research projects and reciprocal exchanges. One of the objectives of such collaborations should be building a shared understanding of each other's priorities for the development of harmonised global norms and standards. Fifth, develop and disseminate biosecurity handbooks or other communication toolkits to biosafety professionals and members of institutional biosafety/bioethics committees to strengthen biosecurity governance at the institute level. Lastly, institutionalise engagement with community and civil society actors for bio-surveillance and biorisk management.

Conclusion

Effective biosecurity governance depends on a combination of legal and regulatory systems, institutions, resources and culture that could respond to the rapidly evolving biorisk landscape. There are several challenges to the existing governance practices, including limited and uneven awareness and funding, a lack of oversight and compliance mechanisms, and limited North-South and South-South collaborations. Further, biosecurity-related laboratory capacity, technical norms and standards and funding streams are concentrated in a small number of Western countries. The uneven distribution of risks and capacity and diverse cultural norms and government systems necessitate the development of non-Western narratives for biosecurity governance. The varied governance models mentioned in the case studies show diverse practices, which should be mainstreamed for developing a more balanced global biosecurity governance.

Several global instruments often presume similar challenges and potential solutions, while States and communities have varied histories, priorities, institutional structures and existing practices. This could result in regulatory gaps, limited cooperation and legitimacy deficits. Here, BioGovernance Commons could play a crucial role by facilitating sustained, inclusive dialogues and co-production of norms. The Commons provides space for mutual learning where divergent practices could be shared and builds trust and mutual accountability through repeated multi-stakeholder interactions. Through iterative norm-making, it also helps build relational norms that evolve as risks and technology do. Finally, the Commons framework complements the various technical fixes being developed to solve biosecurity challenges by addressing existing inequities.

References

Africa Centres for Disease Control and Prevention (Africa CDC). 2025a. Africa CDC weekly brief, 26–31 May 2025. https://africacdc.org/news-item/africa-cdc-weekly-brief-26-31-may-2025/. Accessed 19 Mar 2026.

Africa Centres for Disease Control and Prevention (Africa CDC). 2025b. Regional biosafety and biosecurity legal framework for the African union member states. https://africacdc.org/download/regional-biosafety-and-biosecurity-legal-framework-for-the-african-union-member-states/. Accessed 6 Aug 2025.

ASEAN Hub. 2025. ASEAN nations advancing regional strategy to strengthen biosafety & biosecurity. https://aseanhub.org/2025/09/18/asean-nations-advancing-regional-strategy-to-strengthen-biosafety-biosecurity/. Accessed 18 Sept 2025.

Asia Centre for Health Security (ACHS). 2026. Asia centre for health security. https://asia-chs.org/. Accessed 19 Mar 2026.

Association of Southeast Asian Nations (ASEAN). 2024. *ASEAN leaders' declaration on strengthening regional biosafety and biosecurity*. https://asean.org/wp-content/uploads/2024/10/6-ASEAN-Leaders_-Declaration-on-Strengthening-Regional-Biosafety-and-Biosecurity.pdf. Accessed 9 Oct 2024.

Attorney-General's Chambers. 1960. *Internal security act 1960*. https://sso.agc.gov.sg/Act/ISA1960.

Attorney-General's Chambers. 2002. *Terrorism (suppression of financing) act 2002*.https://sso.agc.gov.sg/Act/TSFA2002.

Bielecka-Oder, A. 2018. Safety and security regulations against biological threats. In *Defence against bioterrorism: Methods for prevention and control*, edited by V. Radosavljevic, I. Banjari, and G. Belojevic, 151–176. https://doi.org/10.1007/978-94-024-1263-5_12.

Biotechnology Industry Research Assistance Council (BIRAC). 2024. *India bioeconomy report 2024*. https://birac.nic.in/webcontent/IBER_2024.pdf.

Bipartisan Commission on Biodefense. 2024. *The national blueprint for biodefense: Immediate action needed to defend against biological threats*. April 2024. https://biodefensecommission.org/wp-content/uploads/2024/05/National-Blueprint-for-Biodefense-2024_final_digital.pdf.

BWC Implementation Support Unit. n.d. *Handbook for youth and biosecurity (Nigeria)*. https://bwcimplementation.org/sites/default/files/resource/Nigeria_HandbookYouth4Bio.pdf.

C.K. 2024. How well-funded is biosecurity philanthropy?. https://forum.effectivealtruism.org/posts/pnincG5vW8Far8Ggg/how-well-funded-is-biosecurity-philanthropy. Accessed 5 Apr 2024.

China Daily. 2021. Biosecurity law goes into effect. https://english.mee.gov.cn/News_service/media_news/202104/t20210416_828952.shtml. Accessed 16 Apr 2021.

Congressional Research Service. 2024. U.S. biodefense policy: Overview and issues for congress. https://www.congress.gov/crs_external_products/R/HTML/R48155.web.html.

Dale-Huang, Alexis, and Nathan Beauchamp-Mustafaga. 2024. Chinese military thinking at the crossroads of biological security, biotechnology, and global health. https://www.nbr.org/publication/chinese-military-thinking-at-the-crossroads-of-biological-security-biotechnology-and-global-health/.

Department of Biotechnology (DBT), Government of India. 2026. India bioscience knowledge portal. https://ibkp.dbtindia.gov.in.

Department of Biotechnology, Ministry of Science and Technology. n.d. Review committee on genetic manipulation (RCGM): Secretariat. https://ibkp.dbtindia.gov.in/Content/Commitee.

Directorate General of Foreign Trade. 2018. *Guidelines for special chemicals, organisms, materials, equipment and technologies (SCOMET)*. https://ibkp.dbtindia.gov.in/DBT_Content_Test/CMS/Guidelines/20181115135754468_Export%20of%20SCOMET%20guidelines.pdf.

European Biosafety Association. 2026. European biosafety and biosecurity legislation. https://www.ebsaweb.eu/biosafety/biosecurity-resources/european-biosafety-/-biosecurity-legislation.

Evans, Sam Weiss, Piers Millett, James Revill, and Edward Gaynor. 2020. Embrace experimentation in biosecurity governance. *Science* 368:138–140. https://doi.org/10.1126/science.aba2932.

Federal Select Agent Program. 2025. Home page. https://www.selectagents.gov/index.htm.

Green, Shannon L., and Skandan Ananthasekar. 2023. Review: The UK government's updated biological security strategy. June 28, 2023. https://councilonstrategicrisks.org/2023/06/28/review-the-u-k-governments-updated-biological-security-strategy/.

Home Team Science and Technology Agency (HTX), Singapore. 2026. CBRNE expertise. https://www.htx.gov.sg/who-we-are/what-we-do/our-expertise/cbrne.

IndiaBioscience. 2026. Bio-responsibility future leaders programme. https://indiabioscience.org/events/bio-responsibility-future-leaders-programme.

International Biosecurity and Biosafety Initiative for Science (IBBIS). 2025. IBBIS welcomes EU commitment to biosecurity in upcoming biotech act. https://ibbis.bio/ibbis-welcomes-eu-commitment-to-biosecurity-in-upcoming-biotech-act/.

International Gene Synthesis Consortium (IGSC). 2026. International gene synthesis consortium. https://genesynthesisconsortium.org.

Justen, Nils, Nathan Calvin, and Andrew Weber. 2023. *Battling catastrophic biological risks*. https://councilonstrategicrisks.org/wp-content/uploads/2023/05/BattlingCatastrophicBioThreats.pdf.

Lee, Chloe. 2023. A write-up on the biosecurity landscape in China. https://forum.effectivealtruism.org/posts/Lw8bwJvtCo6ssqEKa/a-write-up-on-the-biosecurity-landscape-in-china.

Liang, Huigang, Cui Huang, Xiaoli Zhu, and Zhiming Yuan. 2021. Significance of and outlook for the biosecurity law of the people's republic of China. *Journal of Biosafety and Biosecurity* 3 (1): 46–50. https://doi.org/10.1016/j.jobb.2021.06.004.

Linkous, B., R. N. Burnette, and S. Dittrich. 2021. Biodefense promotes biosecurity through threat reduction programs and global health security. https://doi.org/10.1007/978-3-030-69464-7_4.

Maruta, Talkmore, et al. 2023. Regional approach to strengthening biosafety and biosecurity systems in Africa. https://doi.org/10.1080/23779497.2023.2257766.

Min, K., Y. Zhang, J. Liu, et al. 2025. China's biosafety/biosecurity governance. *Frontiers of Medicine* 19:871–878. https://doi.org/10.1007/s11684-025-1158-y.

Ministry of External Affairs, Government of India. 2025. International conference on "50 years of BWC." https://www.mea.gov.in/press-releases.htm?dtl%2F40353%2FInternational_conference_on_50_Years_of_BWC_Strengthening_Biosecurity_for_the_Global_South_December_01__02_2025=.

Ministry of Foreign Affairs of the People's Republic of China. 2024. The biological weapons convention. https://www.mfa.gov.cn/eng/wjb/zzjg_663340/jks_665232/kjlc_665236/shwq_665244/202406/t20240606_11405160.html.

Ministry of Health, Singapore. 2005. *Biological agents and toxins act*. https://inetapps.nus.edu.sg/osh/portal/general_safety/legal_pdf/Biological_Agent_Toxin_Act.pdf.

Morrison, Stephen J., and Michaela Simoneau. 2023. Eight common-sense actions for biosafety and biosecurity. https://www.csis.org/analysis/eight-commonsense-actions-biosafety-and-biosecurity.

Muluneh, Atalay Goshu, et al. 2025. Mapping biosafety level 3 (BSL-3) and BSL-4 laboratories. https://doi.org/10.1007/s10389-025-02492-3.

Namdeo, Suryesh Kumar, and Chandan G. N. 2025. Scientometric analysis of research on deliberate biosecurity threats. https://doi.org/10.1016/j.jobb.2025.04.002.

Namdeo, Suryesh Kumar, Fatima Aziz, and Michael Scott Friedman. 2026. Developing Multistakeholder Engagement to Strengthen Biosecurity. Health Securit 24 (2): 2026. Accessed 27 Mar 2026. https://doi.org/10.1177/23265094261438869

Namdeo, Suryesh Kumar, and Joy Y. Zhang. 2024. Rethinking science diplomacy and global biosecurity. https://doi.org/10.1093/ia/iiae187.

National Academies of Sciences, Engineering, and Medicine. 2020. Key roles for the department of defense in biological threat reduction. https://doi.org/10.17226/25681.

National Biosafety Management Agency (NBMA), Nigeria. 2022. *National biosecurity policy and action plan 2022–2026*. https://nbma.gov.ng/wp-content/uploads/2022/02/National-Biosecurity-Policy-and-Action-Plan-2022-2026.pdf.

National Institute of Allergy and Infectious Diseases (NIAID). 2026. Biocontainment research facilities. https://www.niaid.nih.gov/research/biocontainment-research-facilities.

Novossiolova, Tatyana A., et al. 2021. The vital importance of a web of prevention. https://doi.org/10.1186/s42522-021-00049-4.

NTI | bio. 2024a. AIxBio global forum. https://www.nti.org/about/programs-projects/project/aixbio-global-forum/.

NTI | bio. 2024b. Exploring AI-biosecurity governance in the global south. https://www.nti.org/risky-business/exploring-ai-biosecurity-governance-in-the-global-south/.

NTI | bio. 2025. NTI advances biotechnology governance framework. https://www.nti.org/news/nti-advances-biotechnology-governance-framework-to-reduce-emerging-biological-risks/.

Office of the Principal Scientific Adviser to the Government of India. 2026. One health mission. https://www.psa.gov.in/oneHealthMission.

Prouse, Carolyn, and Mohammed Rafi Arefin. 2025. Geographies of biosecurity. https://doi.org/10.1177/03091325251358410.

Qiushi (求是). 2023. China contributes its part to global biosecurity governance. https://en.qstheory.cn/2023-08/11/c_909910.htm.

Rawluk, A., R. Beilin, and S. Lavau. 2021. Enacting shared responsibility in biosecurity governance. https://doi.org/10.5751/ES-12368-260218.

Reed, M. S., and R. Curzon. 2015. Stakeholder mapping for the governance of biosecurity. https://doi.org/10.1080/1943815X.2014.975723.

Responsible AI x Biodesign Community. 2024. Community values, guiding principles, and commitments. https://responsiblebiodesign.ai/#about.

Revill, James. 2024. How the biological weapons convention could verify treaty compliance. https://thebulletin.org/2024/03/how-the-biological-weapons-convention-could-verify-treaty-compliance/.

Singapore Customs. 2026. Strategic goods control act. https://www.customs.gov.sg/businesses/acts-and-subsidiary-legislation/strategic-goods-control-act/.

Tan, Hao Yi. 2025. Overview of Singapore's one health strategy. https://doi.org/10.3390/zoonoticdis5010002.

The White House. 2022. *National biodefense strategy and implementation plan*. https://bidenwhitehouse.archives.gov/wp-content/uploads/2022/10/National-Biodefense-Strategy-and-Implementation-Plan-Final.pdf.

Trump, Benjamin D., et al. 2021. Promoting effective biosecurity governance. https://doi.org/10.1007/978-94-024-2086-9_14.

U.S. Department of Health and Human Services. 2024. *U.S. government policy for oversight of DURC and PEPP*. https://aspr.hhs.gov/S3/Documents/USG-Policy-for-Oversight-of-DURC-and-PEPP-May2024-508.pdf.

U.S. Department of State. 2026. Office of cooperative threat reduction. https://www.state.gov/key-topics-office-of-cooperative-threat-reduction.

UK Government. 2025. *UK biological security strategy: Implementation report.* https://www.gov.uk/government/publications/uk-biological-security-strategy-implementation-report-june-2023-june-2025/uk-biological-security-strategy-implementation-report-june-2023-june-2025-html.
United Nations Office for Disarmament Affairs (UNODA). 2026. China and Lao PDR co-host first regional workshop. https://disarmament.unoda.org/en/updates/china-and-lao-pdr-co-hosted-first-regional-workshop-implementing-biological-weapons.
Vennis, Iris M., et al. 2022. Complementarity of international instruments in the field of biosecurity. https://doi.org/10.3389/fpubh.2022.894389.
Wang, Leifan, Jie Song, and Weiwen Zhang. 2021. Tianjin biosecurity guidelines. https://doi.org/10.1016/j.jobb.2021.08.001.
Xiao, Yang, Xue Yang, and Tian Deqiao. 2023. China's biosafety/biosecurity concept and governance system. *Contemporary International Relations* 33 (3): 97–120.
Zhang, Joy Y., Sonia Ben Ouagrham-Gormley, and Kathleen M. Vogel. 2022. Creating common ground with Chinese researchers. *Issues in Science and Technology* 38 (4): 45–48. https://www.jstor.org/stable/27161960.

Suryesh Kumar Namdeo is a Senior Research Analyst at the Indian Institute of Science, Bangalore. He is also a member of the Indian National Young Academy of Science (INYAS). Suryesh leads policy research and engagement projects on bio-responsibility and science diplomacy. He also runs the Bio-responsibility Future Leaders Programme (BFLP) at the Indian Institute of Science that aims to train, upskill and empower early career researchers to address key policy challenges in biosecurity, biosafety and bioethics in an integrated manner.

He is currently a member of the Advisory Board of the International Biosecurity and Biosafety Initiative for Science (IBBIS) and has contributed to its efforts including the Global DNA Synthesis Map project. Previously, he was a Fellow at the Asia Centre for Health Security, National University of Singapore. He has also served as the science and technology consultant for the Biological Weapons Convention (BWC) at the United Nations Office for Disarmament Affairs in Geneva. Here, his responsibilities included facilitating science and technology advice to diplomats and international civil servants, and preparing a multi-stakeholder engagement strategy for the BWC. He has co-authored the 'Seoul Youth Declaration for Disarmament and Non-proliferation' and the 'Youth Declaration for Biosecurity' in efforts coordinated by the United Nations Office for Disarmament Affairs. He earlier served as the programme officer for the Science, Technology and Innovation Policy Fellowship Programme of the Government of India, where he coordinated the programme, built partnerships and conducted capacity building activities. He also held the role of Director of Outreach and Engagement for the Journal of Science Policy and Governance (JSPG). He holds a PhD in Molecular Biology from the Max Planck Institute for Biology, Germany and a professional certification in biorisk management from the International Federation of Biological Safety Associations (IFBA).

Chapter 8
Bilateral Bio-Governance of Traditional Chinese Medicine: A Case Study of China-Malaysia Initiatives

Mofei Lyu and Hon-Ngen Fung

Abstract The global expansion of Traditional Chinese Medicine (TCM) marks a significant transformation in contemporary global health governance, moving beyond cultural exchange or commercial circulation toward the institutionalization of a state-backed, standardized medical system. This article examines bilateral bio-governance through a case study of China–Malaysia cooperation on TCM, conceptualized within the framework of the Biogovernance Commons. This chapter argues that China's international promotion of TCM constitutes a form of strategic knowledge leadership, embedded in national development planning and supported by a powerful sociotechnical imaginary that links medical modernization, public health, and global responsibility. The analysis identifies four interrelated axes of successful biogovernance: epistemic legitimation, regulatory alignment, knowledge transmission, and discursive amplification. This initiative goes beyond cultural exchange or trade, establishing a robust, state-backed system that seeks to institutionalize a standardized, high-quality structure of medical knowledge and practice across international boundaries. The successful harmonization of standardized TCM with national regulatory regimes and global health practices offers an innovative, complementary dimension to the global health landscape. This chapter analyses the Sino-Malaysian partnership, focusing specifically on how dynamic policy mechanisms successfully drive institutional integration, establish comprehensive regulatory standards, and facilitate *mutual institutional development* within the crucial domain of the Biogovernance Commons.

Keywords Traditional Chinese medicine · Biogovernance · Traditional knowledge

---

M. Lyu
Department of Science and Technology Studies, Faculty of Science, Universiti Malaya, Kuala Lumpur, Malaysia

H.-N. Fung (✉)
Department of Science and Technology Studies, Faculty of Science, Universiti Malaya, Kuala Lumpur, Malaysia
e-mail: kennethfung@um.edu.my

J. Y. Zhang et al. (eds.), *BioGovernance Commons and China's Role in Global Risk Governance*, Emerging Globalities and Civilizational Perspectives,
https://doi.org/10.1007/978-3-032-32198-5_8

Introduction

The Institutionalization of Strategic Knowledge Leadership

Strategic knowledge leadership, in this context, describes the deliberate, state-led attempt to shape how medical knowledge—specifically modernized TCM emanating from China—is efficiently standardized, rigorously validated, and professionally diffused globally (Basner 2024; Wang et al. 2016). China's dedication to advancing TCM internationally is not peripheral; it is fundamentally integrated into national strategic planning and contributes significantly to the global public good. Key documents, such as the *Outline of Strategic Plan on the Development of Traditional Chinese Medicine (2016–2030)*, explicitly elevate the global promotion of TCM to a national strategic priority, positioning it as an indispensable component of public health development, national soft power, and technological modernization (Basner 2024; Ministry of Health of the People's Republic of China 2016). This strategic positioning successfully transforms TCM from a merely therapeutic system into a strategic *cultural technology* and a powerful tool for *experience sharing* (Basner 2024).

This elevation is driven by a powerful *sociotechnical imaginary* (Winkler 2023). A sociotechnical imaginary constitutes a collective, future-oriented vision of scientific and technological progress that is understood to be inextricably linked to the public good, collective destiny, and political identity of a community (Winkler 2023). It is the "dream" of how a specific technology or science will improve society, define a nation's identity, and solve global problems. For the People's Republic of China, the global recognition of TCM signifies a modern future where sophisticated Chinese medical knowledge is a universally accepted contributor to global health and well-being, simultaneously projecting national influence and demonstrating a distinctive, modernized cultural pride (Basner 2024; Ministry of Health of the People's Republic of China 2016; Winkler 2023). The goal is to create a future where Chinese medical knowledge is a universally accepted contributor to global health, projecting national pride and "soft power." This strategic framework successfully establishes a stable basis for *mutual institutional development* (Wang et al. 2016). China's strategic policy provides the necessary momentum and models for partner nations, such as Malaysia, enabling them to rapidly modernize and align their own legislative and institutional frameworks to embrace these high international standards (Ministry of Health of the People's Republic of China 2016).

Defining the Collaborative Biogovernance Commons

The Biogovernance Commons is understood here as the shared, dynamic arena encompassing health resources, diverse knowledge systems, and the legal-regulatory infrastructures established to manage and harmonize these domains (Law and

Cloatre 2019; Renn 2008, as cited in Law and Cloatre 2019). Within this collaborative space, the Sino-Malaysian partnership provides an exemplary model for *proactive risk governance*, particularly at the critical interface where traditional, culturally rich knowledge systems are modernized and integrated into modern state regulation (Law and Cloatre 2019).

The successful collaboration is evident in the regulatory progress achieved by Malaysia. Following global calls for integration, such as the *WHO Traditional Medicine Strategy 2014–2023* (World Health Organization 2013), the strategic support from China facilitated Malaysia's integration process, effectively mitigating regulatory challenges. By sharing its robust, nationally standardized model of TCM, China provides a proven blueprint for regulatory systems, allowing the host country to address complexities effectively (Law and Cloatre 2019; Ministry of Health of the People's Republic of China 2016). Malaysia's institutional achievement, marked by the enforcement of the *Traditional and Complementary Medicine Act 2016 (Act 775)*, requires TCM to comply with state-based quality assurance techniques, including professional certification and evidence standards, which successfully align TCM with the best practices of modern conventional medicine (Department of Health Malaysia 2016; Law and Cloatre 2019; Shamsul and Said 2024). This positive adaptation process generates substantial internal capacity in Malaysia, demonstrating how China's strategic policy push serves as a crucial vector for efficient institutional reform and collaborative complexity management within the global Biogovernance Commons (Ministry of Health of the People's Republic of China 2016).

Policy Temporalities, Standardization, and Institutional Alignment

A critical analysis of policy document production and release schedules reveals a planned, cyclical rhythm that governs the intensity and tempo of TCM internationalization. This rhythm is crucial; it serves as a mechanism for targeted *strategic steering*, effectively orchestrating institutional reform and driving *efficient institutional alignment* in partner nations (Currie and Seddon 2022).

Strategic Rhythms and Proactive Policy Synchronization

An examination of official policy documents related to TCM since 2011 demonstrates a deliberate, non-monotonic, and distinctly periodic fluctuation in publication volume, following a discernible "two-peak, valley-in-between" trajectory (Ministry of Health of the People's Republic of China 2016). This cyclical pattern directly reflects the centralized, top-down efficiency of Chinese national strategic planning, embodying a proactive *biopolitical temporality*—the state's deliberate management and extension of its influence through time to ensure optimal health outcomes globally (Ministry of Health of the People's Republic of China 2016).

The major policy surges observed were strategically concentrated around the years 2016 and 2021 (Ministry of Health of the People's Republic of China 2016). The 2016 peak coincided directly with the issuance of the *Outline of the Strategic Plan on the Development of Traditional Chinese Medicine (2016–2030)*, which successfully elevated TCM internationalization to a long-term, high-level political mandate (Ministry of Health of the People's Republic of China 2016). The subsequent peak around 2021 aligned with the commencement of the 14th Five-Year Plan period, signaling the effective integration of TCM advancement into China's medium-term economic and public health development strategy (Ministry of Health of the People's Republic of China 2016). This concentration of high-level policy release during specific, politically determined windows creates strategic opportunities to facilitate rapid *normative alignment* and accelerate the adoption of international best practices in partner nations (Currie and Seddon 2022).

Institutional theory highlights that organizations or states that collaborate closely, or those benefiting from strategic resource transfer, often enthusiastically adopt similar organizational or legislative structures to enhance legitimacy and access to promised resources (Currie and Seddon 2022). The successful policy synchronization is evident in Malaysia's introduction of its foundational regulatory framework, the *Traditional and Complementary Medicine Act* (Act 775), in 2016 (Department of Health Malaysia 2016)—timed perfectly during a major Chinese policy peak. This swift regulatory action demonstrates Malaysia's decisive step to align its legal framework and efficiently access the strategic, economic, and political benefits outlined in China's ambitious internationalization efforts (Ministry of Health of the People's Republic of China 2016). This proactive prioritization of institutional convergence ensures rapid legislative maturity and accelerates local regulatory preparedness (Park et al. 2022; Shamsul and Said 2024).

From Principled Initiatives to Mechanical Docking: The Techno-Legal Advance

The evolution of China–Malaysia TCM cooperation policy demonstrates a commitment to operational excellence, shifting its governance goal from broad "principled initiatives" to concrete requirements for "mechanism docking" (Ministry of Health of the People's Republic of China 2016). While early policy steps involved useful bilateral memoranda of understanding (MoUs) to signal political alignment (Embassy of the People's Republic of China in Malaysia 2022), recent trends emphasize specific, verifiable, and technical procedures, such as efficiently streamlining Halal certification processes, standardizing drug registration approvals, and unifying technical terminology (Syed Ismail al-Qudsy et al. 2025). This detailed focus successfully eliminates tangible institutional barriers to trade and practice (Syed Ismail al-Qudsy et al. 2025).

This conceptual transition signifies a critical *techno-legal advance* in TCM governance (Law and Cloatre 2019). The primary policy challenge has successfully shifted from achieving abstract political alignment to attaining concrete *technical alignment* between two distinct regulatory apparatuses. This tendency underscores the profound influence of *law and biomedicine* as harmonizing forces that efficiently impose necessary quality standards (e.g., Pharmacopoeia standards, Good Manufacturing Practice) onto traditional systems (Law and Cloatre 2019; Wang et al. 2016). By centering policy discussions on tangible procedures like registration, technical standards, labeling, and certificates, the policy process successfully reinforces the necessity for TCM to translate its knowledge claims into a globally recognized technical language of regulatory science, thus enhancing product safety and reliability (Law and Cloatre 2019; Wang et al. 2016). The sustained pursuit of standardization is fundamentally the pursuit of *institutional legitimation* within a responsible global technoscientific regime (Law and Cloatre 2019).

Thematic Policy Analysis: Axes of Standardization and Institutional Success

A systematic policy analysis of documents guiding China-Malaysia cooperation reveals core thematic axes of successful engagement (Ministry of Health of the People's Republic of China 2016). These axes, derived from the functional priorities of the policy texts, encapsulate the primary areas of collaboration and mutual development within the Biogovernance Commons. The policy discourse clusters around eight specific areas, which can be synthesized into four overarching thematic axes: Epistemic Legitimation, Regulatory Alignment, Knowledge Transmission, and Discursive Amplification (Table 8.1).

Epistemic Legitimation and Institutional Integration in Health Systems

Policy discourse related to public health integration consistently prioritizes the functional and systemic role of TCM within Malaysia's existing healthcare architecture (Ministry of Health of the People's Republic of China 2016). Key policy concepts such as *intervention*, *effectiveness*, *availability*, and *diagnosis and treatment* highlight a proactive focus on the tangible utility and successful placement of TCM services (Ministry of Health of the People's Republic of China 2016). Malaysia's adoption of the *Traditional and Complementary Medicine Act 2016* formally grants high legal status to TCM and practitioners, marking a significant

Table 8.1 Thematic axes in Sino-Malaysian TCM policy discourse

Thematic axis	Policy focus (institutional locus)	Key policy manifestations and indicators	STS-informed success
Epistemic legitimation	Public health system integration	Health care, diagnosis and treatment, hospitals, availability, intervention, effectiveness, rehabilitation, medicinal materials, perception, pain (Ministry of Health of the People's Republic of China 2016).	Successfully defining "scientific merit" and achieving comprehensive regulatory legitimacy for TCM, facilitating its *institutional integration* (Law and Cloatre 2019).
Regulatory alignment	Trade, compliance, and market entry	Certificate, approval, label, poison, market launch, export, trade barriers, inspection, comply, sign/record, application, amendment, formulation (Ministry of Health of the People's Republic of China 2016).	Proactively resolving institutional complexities and minimizing *regulatory fragmentation* through dedicated docking mechanisms (Currie and Seddon 2022; Ministry of Health of the People's Republic of China 2016).
Knowledge transmission	Education and professional standards	Professionalization, diploma, certificate, academic, bachelor, accreditation, federation, apprentice, workshop (Ministry of Health of the People's Republic of China 2016).	Effectively managing the *cultural translation* and standardization of professional expertise, ensuring fidelity and high quality (Sanders 2000; Wang et al. 2016).
Discursive amplification	Media, public trust, and soft power	Marketing, interview, advertisement, news, author, session, prescription, Malaysia, traditional medicine, health preservation, health care, medicine (Ministry of Health of the People's Republic of China 2016).	Strategically shaping the *sociotechnical imaginary* of TCM, building public confidence, and providing crucial information during *infodemics* (Wang and Wang 2025; Winkler 2023).

institutional achievement (Department of Health Malaysia 2016; Shamsul and Said 2024).

The policy discourse successfully utilizes evaluation metrics—such as assessing clinical *effectiveness* and defining appropriate *intervention* protocols—that are rooted in global standards, which is necessary for rigorous quality assurance (Ministry of Health of the People's Republic of China 2016). This regulatory requirement facilitates critical *institutional integration*, confirming that legal and scientific standards converge to define and recognize legitimate traditional medicine (Law and Cloatre 2019). The legal acceptance of TCM in Malaysia thus signifies an advanced process of *professionalization* and *institutional legitimation* (Law and Cloatre 2019). TCM practitioners are proactively adapting their traditional knowledge claims and therapeutic practices to fit robust national evidentiary requirements, ensuring the system's knowledge base aligns with the crucial strategic standardization goals promoted by China (Law and Cloatre 2019).

Furthermore, the focus on *availability* and *hospitals* (Ministry of Health of the People's Republic of China 2016) confirms a commitment to accessibility. State cooperation strategically facilitates the flow of TCM biocapital, supporting the provision of subsidized TCM services in government settings (Shamsul and Said 2024). This strategic economic support ensures that the institutional integration, while achieving legal recognition, also promotes broader accessibility for the population, rooted in a positive, market-responsive governance model.

Regulatory Alignment and Institutional Harmonization

The second thematic axis, concerning regulation and trade, focuses heavily on achieving the technical and logistical requirements necessary for seamless market access (Ministry of Health of the People's Republic of China 2016). High-frequency policy words in this area include *certificate*, *label*, *comply*, *application*, *market launch*, and *trade barriers* (Ministry of Health of the People's Republic of China 2016). The prevalence of these terms confirms that policy aims are actively and successfully addressing systemic complexity and promoting regulatory alignment (Park et al. 2022). Although some regulatory complexities naturally exist in cross-border cooperation, policy aims for proactive "mechanism docking," successfully moving China and Malaysia toward an integrated, mutually recognized TCM regulatory and certification system (Malaysia National Pharmaceutical Regulatory Agency 2023).

This institutional coherence is achieved by policy mechanisms designed to overcome *institutional complexities* (Ministry of Health of the People's Republic of China 2016). Regulatory standards are being harmonized to reduce inconsistencies and lower transactional costs (Ministry of Health of the People's Republic of China 2016; Tan and Omar 2022). The successful collaboration between Chinese standardization efforts (like the Chinese Pharmacopoeia) and Malaysian local requirements, such as Halal certification, showcases a dedication to cultural responsiveness (Syed Ismail al-Qudsy et al. 2025). Companies are empowered to meet both rigorous Chinese quality standards and comply with Malaysia's culturally specific Halal certification requirements, a testament to effective policy adaptation (Department of Standards Malaysia 2012; Syed Ismail al-Qudsy et al. 2025). This regulatory diligence, backed by China's national strategic policy (Ministry of Health of the People's Republic of China 2016), proves highly effective in promoting quality and establishing *mutual recognition*, the foundational goal of a mature Biogovernance Commons (Malaysia National Pharmaceutical Regulatory Agency 2023).

Knowledge Transmission, Education, and Professional Standardization

The policy dimension successfully interfaces with the education helix, establishing the necessary institutional infrastructure for professional knowledge transfer and legitimation (Ministry of Health of the People's Republic of China 2016). Policy topics in this area focus intensely on *professionalization*, *diploma*, *certificate*, *academic*, and *accreditation* (Ministry of Health of the People's Republic of China 2016). Policy interventions proactively provide institutional support for bilateral education agreements, degree recognition, and joint educational programs (Wong et al. 2019). A prime example is the successful TCM undergraduate program launched by Management & Science University of Malaysia and Henan University of Chinese Medicine (MSU and HUCM 2023).

The policy function in this axis acts as the institutional guarantor, ensuring that TCM professionals trained in either jurisdiction meet the necessary high standards and cultural requirements of both nations (Wong et al. 2019). This policy-education linkage is central to the positive process of *cultural translation*, which is the necessary process of adapting existing health knowledge systems to fit local cultural norms while maintaining the functional fidelity and "scientific merit" of the original system (Sanders 2000). A major success in this area is the integration of Halal certification for TCM products. This allows the medicine to meet both rigorous Chinese quality standards and Malaysian religious requirements. The robust emphasis on formal accreditation and professional certification successfully serves as a mechanism for state-led quality control over the consistency and high fidelity of the transmitted knowledge (Ministry of Health of the People's Republic of China 2016). This process is crucial for establishing and stabilizing the perceived safety and efficacy of TCM globally, especially in a contemporary environment where public trust is challenged (Wang and Wang 2025). Policy thus clearly defines the acceptable *epistemological foundation* for TCM practice, ensuring that the transmitted knowledge fully aligns with the strategic and political goals of both cooperating states (Ministry of Health of the People's Republic of China 2016). Epistemological collaboration refers to different ways of knowing—like traditional wisdom and modern biomedicine—working together to create a shared set of rules. This results in shared tools like the Malaysian herbal pharmacopeia, which sets official standards for what constitutes "safe" and "effective" medicine.

Discursive Amplification, Media, and Societal Confidence

The fourth thematic axis, Discursive Amplification, examines how policy and state objectives positively utilize the media sphere to shape public perception and acceptance of TCM (Ministry of Health of the People's Republic of China 2016). This dimension employs keywords such as *marketing*, *interview*, *advertisement*, *news*, and *health preservation* (Ministry of Health of the People's Republic of China 2016). This thematic focus confirms that policy-making actively involves the successful construction and dissemination of a favorable *sociotechnical imaginary* of TCM within the host country (Winkler 2023).

Official communication and governmental policy priorities directly influence the media helix by defining which TCM topics receive institutional attention and funding for communication projects (Ministry of Health of the People's Republic of China 2016). The official support for TCM, particularly its elevation to a national strategic level in China, successfully translates into frequent positive media coverage, thereby significantly enhancing public awareness and societal confidence (Ministry of Health of the People's Republic of China 2016). The policy–media loop is particularly valuable during periods of global crisis. For example, official statements from Chinese foreign ministry spokespersons publicly emphasizing TCM's strategic status and its efficacy in COVID-19 prevention and treatment (Zhao 2021) were effectively amplified by media channels to reinforce public confidence and legitimize the system (Ministry of Health of the People's Republic of China 2016). This loop demonstrates policy acting as a proactive mechanism for crisis legitimation and the strategic generation of public trust, essential for counteracting the destabilizing effects of *infodemics* (Law and Cloatre 2019; Wang and Wang 2025). By successfully managing societal expectations, policy ensures the socio-cultural components of the biogovernance commons remain stable and fully receptive to the foreign knowledge system (Ministry of Health of the People's Republic of China 2016).

Strategic Policy Discourse Analysis: Leadership, Alignment, and Mutual Benefit

A deeper Strategic Policy Discourse Analysis (SPDA) confirms that the implementation of the Sino-Malaysian TCM cooperation policy is structurally supported by a clear division of roles, leading to high synergy and a productive relationship (Basner 2024; Currie and Seddon 2022).

China's Leadership in Experience Sharing and Institutional Support

An examination of the political rhetoric embedded in official bilateral communication reveals a complementary *role alignment* in policy discourse (Basner 2024). Official Chinese statements productively employ discursive strategies that assert a guiding, directional role, utilizing phrases such as "China leads... cooperation" or "exports experience" (Basner 2024). Conversely, Malaysian policy documents and official statements adopt language reflecting a supportive, receptive, and adaptable role, utilizing terms like "should learn from" and "receive support" (Basner 2024; Embassy of the People's Republic of China in Malaysia 2022).

This constructive "we–they" framework successfully institutionalizes a stable, *asymmetric biopower* relationship where leadership is welcome, and resources are optimized (Law and Cloatre 2019). This clear hierarchy allows China's national strategic objectives—primarily soft power projection, global market expansion, and the standardization of its traditional medical knowledge—to effectively steer the bilateral cooperation agenda (Ministry of Health of the People's Republic of China 2016). By framing the relationship as one of strategic support and experience transfer, the policy discourse successfully *mobilizes* necessary economic dependencies and political power relations for mutual benefit (Currie and Seddon 2022). This mobilization promotes critical discussion regarding equitable resource distribution, genuine local ownership of standards, local accessibility, and the foundational aspects of health justice within the partnership (Shamsul and Said 2024). The policy thus functions as a medium for institutionalizing a systemic *epistemic collaboration* that serves as a robust basis for the equitable *co-production* of shared medical and regulatory standards (Law and Cloatre 2019; Ministry of Health of the People's Republic of China 2016).

Policy Orientation, Consistency, and Successful Implementation

A primary source of high policy synergy stems from a fundamental consistency in the core strategic orientations of the two nations (Park et al. 2022). China's approach to TCM promotion is deeply strategic and geopolitical, integrating it into national soft power projection and major national public health and development blueprints, providing a clear path forward (Ministry of Health of the People's Republic of China 2016). Malaysia, in parallel, views TCM primarily through the lens of national public health needs and bilateral economic cooperation, treating it as a valuable and supplementary component of its existing healthcare system and industrial development portfolio (Park et al. 2022).

This functional alignment means that both nations enthusiastically agree on high-level policy goals (e.g., "promote exchanges"), with their strategic interests converging efficiently, leading to rapid implementation at the operational level (Park et al. 2022). A critical indicator of this underlying orientation success is the strategic use of high-level, abstract policy language, such as general calls to merely "promote exchanges" (Tan and Omar 2022). This strategic ambiguity serves a crucial political function: it allows both parties to publicly affirm successful cooperation and maintain political harmony while simultaneously providing a flexible framework for detailed, concrete commitments that allow adaptation to local needs (Tan and Omar 2022). This flexibility creates significant *implementation certainty* for actors in the other helices (e.g., industry, education, media), facilitating the development of effective, coordinated bilateral development (Park et al. 2022).

Policy Intersections and Cross-Cultural Governance Success

Within the Quadruple Helix collaborative mechanism (Policy, Education, Media, Economy), the policy dimension successfully operates as the critical "institutional foundation" (Ministry of Health of the People's Republic of China 2016), expertly orchestrating cross-sectoral integration and facilitating necessary cultural adaptation in the host country (Currie and Seddon 2022; Sanders 2000). This institutional function reveals the state's capacity to both enable the successful flow of *biocapital* and manage complex socio-cultural demands required for high legitimation.

Policy and the Economic Helix: Optimizing Biocapital Flow

Policy directly structures the framework for robust economic cooperation by defining clear trade agreements, stable market access regulations, certification standards, and industrial incentives (Ministry of Health of the People's Republic of China 2016). Governmental policy interventions, such as negotiating lower tariffs or simplifying registration processes, directly facilitate the circulation of high-quality TCM products and services as *biocapital*—knowledge and products leveraged for mutual economic gain—into the Malaysian market (Malaysia National Pharmaceutical Regulatory Agency 2023; Wang et al. 2016).

This policy–economy linkage is dynamic and reciprocal. Economic performance feedback, derived from the successful operational outcomes and practical implementation experience of Chinese TCM enterprises in Malaysia, consistently serves as a prompt for necessary policy revisions (World Health Organization 2013). Successful business models and the achievement of regulatory compliance motivate governments to refine regulations or introduce new measures to proactively support industrial growth (World Health Organization 2013). This feedback mechanism demonstrates the governmental structure actively optimizing the market function, reinforcing the *strategic market-responsive* dimensions of the cooperation by prioritizing commercial efficacy, trade compliance, and the reduction of market barriers for the collective good (Currie and Seddon 2022). The core aim of this intersection is to achieve a stable *techno-economic regime* that allows TCM biocapital to flow efficiently and safely (Wang et al. 2016).

Policy and Cross-Cultural Translation: The Halal Case Study

A vital function of the policy helix is its successful management of *cross-cultural translation* (Sanders 2000). For TCM to achieve deep integration and broad societal acceptance in a multireligious and multicultural society like Malaysia, policy frameworks must adapt regulatory policies to meet specific religious and cultural requirements. The necessity of establishing clear, accessible mechanisms for Halal certification is a prime success example, critical for ensuring wide acceptance of TCM products (Syed Ismail al-Qudsy et al. 2025).

The bilateral agreement to jointly draft the Malaysian herbal pharmacopeia (Embassy of the People's Republic of China in Malaysia 2022) illustrates a high level of policy intervention strategically aimed at *institutional adaptation* (Syed Ismail al-Qudsy et al. 2025). The Halal certification requirement highlights policy as a critical site for mediating *positive state–society relations* (Syed Ismail al-Qudsy et al. 2025). By proactively accommodating deeply held cultural and religious needs, the Malaysian state successfully enhances the *normative legitimacy* of TCM (Law and Cloatre 2019), enabling it to be enthusiastically accepted by the broader Malaysian community (Syed Ismail al-Qudsy et al. 2025). In this function, policy actively balances global, standardized (Chinese) norms with deep local, religious contexts (World Health Organization 2013), demonstrating the sophisticated role of governance in successfully stabilizing the socio-cultural components of the biogovernance commons.

Policy, Education, and Institutional Coherence

The policy guidance on education successfully provides accreditation support and ensures high institutional coherence for TCM practice (Wong et al. 2019). The emphasis on formalized qualifications (*professionalization, diploma, certificate, academic*) is strategically designed to upgrade and unify professional standards within the Malaysian TCM sector (Ministry of Health of the People's Republic of China 2016). The legal requirement under the *T&CM Act 2016* for practitioners to complete a 1-year residency, coupled with the formation of the TCM Council, represents a successful state-led effort to impose stronger quality assurance (Shamsul and Said 2024).

Policy, through the mechanism of bilateral educational agreements (e.g., MSU and Henan University of Chinese Medicine joint programs), provides an external, international model for this domestic professionalization drive (MSU and HUCM 2023; Wong et al. 2019). This process successfully induces *normative alignment* (Currie and Seddon 2022), as Malaysian institutions model their standards on established international (Chinese) benchmarks to gain global recognition and domestic legitimacy (MSU and HUCM 2023). The policy thus strategically secures the necessary high-quality human capital pipeline while simultaneously reinforcing the institutional coherence required for a standardized medical system to operate safely and effectively (Ministry of Health of the People's Republic of China 2016).

Conclusion: Achieving Synergy and Structural Success in the Biogovernance Commons

The extensive thematic and discursive analysis confirms that policy, operating as the institutional nucleus of the Quadruple Helix Model, successfully performs a vital dual function in the internationalization of TCM (Ministry of Health of the People's Republic of China 2016). It provides the necessary legal safeguards and institutional platforms required for the cross-cultural integration and high-quality practice of TCM within the host nation's system. Simultaneously, the structure of the policy framework facilitates the successful institutionalization of China's strategic, proactive role in global health—what we have termed *strategic knowledge leadership* (Basner 2024; Ministry of Health of the People's Republic of China 2016).

Structural Successes and the Value of Strategic Alignment

The critical examination of the policy helix reveals persistent structural strengths that enable the achievement of a highly synergistic partnership. The rapid synchronization of Malaysia's institutional framework—exemplified by the *T&CM Act 2016*—with China's major policy cycles (Department of Health Malaysia 2016; Ministry of Health of the People's Republic of China 2016) underscores the policy helix's power in managing institutional alignment and highlights the welcomed structural support extended to the recipient state (Currie and Seddon 2022). The current model successfully prioritizes the objectives of *biocapital* (commercialization and standardization) and *strategic knowledge leadership* (global quality promotion) while simultaneously promoting local health equity and cultural responsiveness (Law and Cloatre 2019; Shamsul and Said 2024).

Four critical structural successes drive the maturation of the Biogovernance Commons in this context:

1. **Policy Orientation Consistency:** The fundamental strategic alignment—China's focus on global quality promotion and Malaysia's complementary healthcare focus—leads to high policy synergy and rapid, concrete implementation (Park et al. 2022).
2. **Regulatory Alignment:** The proactive establishment of harmonized, recognized regulatory and certification systems successfully manages complexity, reducing compliance costs, and facilitating the efficient flow of biocapital and professional services (Malaysia National Pharmaceutical Regulatory Agency 2023; Park et al. 2022).
3. **Epistemic Collaboration in Discourse:** The productive "Leadership–Supportive" discourse structure (Basner 2024) reflects and reinforces the optimized power dynamic, allowing China's established sociotechnical imaginary to

constructively shape the agenda and promote crucial discussions of health quality and shared benefit (Law and Cloatre 2019).

4. **Effective Feedback Mechanisms:** Active participation and timely feedback from non-governmental helices (industry, education, civil society) (Park et al. 2022) ensure that policies are operationally detailed and highly responsive to on-the-ground challenges, creating robust implementation certainty (Park et al. 2022).

Affirming China's Contribution to Global Risk Governance

The policy analysis conducted here, grounded in STS frameworks, offers valuable insights by clarifying how political temporality, discursive strategies, and techno-legal mechanisms successfully interact to construct and legitimize a modernized, high-standard medical-scientific system (Law and Cloatre 2019; Ministry of Health of the People's Republic of China 2016). By highlighting the axes of constructive engagement (Epistemic Legitimation, Regulatory Alignment) and the mechanisms of success (Strategic Leadership, Normative Alignment), this chapter contributes to an appreciative understanding of how effective global health governance is being achieved in the Global South. Ultimately, achieving policy synergy and sustainable integration is successfully driven by China's strategic commitment to standardization and its proactive role in establishing institutional arrangements rooted in mutual development, co-production, and a shared commitment to local health quality, thereby optimizing the terms of access and safety within the Biogovernance Commons (Currie and Seddon 2022; Law and Cloatre 2019; Wang et al. 2016).

References

Basner, A. 2024. *The narration of a 'traditional' medicine in China's political discourse*. Universität zu Köln.

Currie, W., and J. Seddon. 2022. Stakes, positions and logics: An institutional field analysis of cross-border health IT policy. *Journal of Information Technology* 37 (3): 268–285.

Department of Health Malaysia. 2016. *Traditional and complementary medicine act 2016 (Act 775)*. Laws of Malaysia. https://www.moh.gov.my.

Department of Standards Malaysia. 2012. *MS 2424:2012—Halal pharmaceuticals: General guidelines*. Department of Standards Malaysia. https://law.resource.org/pub/my/ibr/ms.2424.2012.pdf.

Embassy of the People's Republic of China in Malaysia. 2022. *China and Malaysia sign Memorandum of Understanding on traditional medicine cooperation*. Ministry of Foreign Affairs of the People's Republic of China. https://www.fmprc.gov.cn.

Law, C. Y., and E. Cloatre. 2019. Law and biomedicine and the making of 'genuine' traditional medicines in global health. *Critical Public Health* 29 (4): 424–434.

Malaysia National Pharmaceutical Regulatory Agency. 2023. *Guidelines for registration of traditional and complementary medicine products*. Ministry of Health Malaysia. https://www.npra.gov.my.

Management & Science University and Henan University of Chinese Medicine. 2023. *Launch of joint institute MSU & HUCM and double degree programme in traditional Chinese medicine*. MSU News.

Ministry of Health of the People's Republic of China. 2016. *Outline of strategic plan on the development of traditional Chinese medicine (2016–2030)*. Beijing: State Council of the People's Republic of China.

Park, J.-E., J. Yi, and O. Kwon. 2022. Twenty years of traditional and complementary medicine regulation and its impact in Malaysia: Achievements and policy lessons. *BMC Health Services Research* 22 (1): 102. https://doi.org/10.1186/s12913-022-07497-2.

Sanders, B. 2000. Cultural adaptation in evidence-based intervention. *Sage*. Publications.

Shamsul, A. B., and S. Said. 2024. An analysis of the rights to health and the traditional complementary medicine (TCM) healthcare services in Malaysia. *International Journal of Academic Research in Business and Social Sciences* 14 (1): 123–145.

Syed Ismail al-Qudsy, S. H., A. Abdul Khalek, and M. Zainuddin. 2025. Governing halal in Malaysia: Innovation towards customer interests and satisfaction. In *The halal industry in Asia: Perspectives from Brunei Darussalam, Malaysia, Japan, Indonesia and China*, 223–242. Springer. https://doi.org/10.1007/978-981-96-0393-0.

Tan, Y. Y., and R. Omar. 2022. Green practices and innovations of traditional Chinese medicine (TCM) industry in Singapore: Idea worth sharing. *Sustainability* 14 (18): 11588. https://doi.org/10.3390/su141811588.

Wang, J., Y. Guo, and G. L. Li. 2016. Current status of standardization of traditional Chinese medicine in China. *Evidence-based Complementary and Alternative Medicine* 2016 (1): 9123103. https://doi.org/10.1155/2016/9123103.

Wang, H. and Wang, Y. 2025. Methodological and conceptual considerations of the TCM evolution. *Proceedings of the National Academy of Science 122(19): e2503785122.*

Winkler, S. C. 2023. *Dreamscapes of modernity: Sociotechnical imaginaries and the global rise of the Chinese state*. Stanford University Press.

Wong, H. F., S. C. Ng, W. T. Tan, J. Liu, X. Lin, S. W. Goh, B. L. Hoo, C. E. Chai, and H. Wang. 2019. Traditional Chinese medicine in Malaysia: A brief historical overview of education and research. *Chinese Medicine and Culture* 2 (3): 114–117.

World Health Organization. 2013. *WHO traditional medicine strategy 2014–2023*. Geneva: World Health Organization. https://www.who.int.

Zhao, L. 2021. *Foreign ministry spokesperson Zhao Lijian's remarks in the TCM and international cooperation forum*. Ministry of Foreign Affairs of China. https://www.fmprc.gov.cn/eng./xw/fyrbt/lxjzh/202405/t20240530_11347013.html.

Mofei Lyu received her master's degree in Science Communication and Public Engagement from the University of Malaya. Her research focuses on the international governance and communication of Traditional Chinese Medicine (TCM), particularly in the context of China–Malaysia cooperation under the Belt and Road Initiative. Drawing on perspectives from Science and Technology Studies (STS), her work examines how policy frameworks, regulatory alignment, and knowledge systems interact to facilitate the cross-border institutionalization of traditional medical knowledge. Her research investigates the policy mechanisms shaping the globalization of TCM through comparative analysis of more than one hundred China–Malaysia policy documents related to TCM cooperation. Using framing analysis, identity construction analysis, and word-frequency analysis, she explores how governments position TCM within policy discourse, how leadership roles are constructed in bilateral cooperation, and what dominant themes emerge in policy narratives. These analyses help identify feasible institutional pathways for China–Malaysia TCM collaboration and provide thematic foundations for examining cross-sector interactions within a Quadruple Helix framework connecting policy, education, industry, and media.

Hon-Ngen Fung is currently a Senior Lecturer at the Department of Science and Technology Studies, Faculty of Science, University Malaya; currently teaching Policy and Management of Sustainable Development, Sociology of Science and Research Methodology. His research expertise focuses on comparative case studies as well as secondary data analysis of policies, research strategies and innovation systems at the sectoral levels for Agriculture (rice, palm oil, etc.), Herbal Medicine, and Renewable Energy (hydrogen, solar, etc.). His published works can be found in journals such as Technological Forecasting & Social Change, Scientometrics, and East Asian Science Technology and Society. Dr Fung is a member of the International Network for Professional Communication of Science and Technology (PCST) and has been an active international speaker on Science Policy, Science Communication and Innovation Systems. He is also currently Deputy Director of the UM Centre for Continuing Education and has experience consulting with the Ministries of Science, Technology and Innovation; Health and Economy in Malaysia.

Chapter 9
The Pessoptimistic Future of the BioGovernance Commons

Joy Y. Zhang

Abstract This concluding chapter reflects on the 'pessoptimistic' future of the BioGovernance Commons as both an empirical initiative and a broader conceptual proposition for global science governance. Drawing on two developments observed during the final stages of editing this volume—restrictions on Chinese participation in Horizon Europe and a case of divergent scholarly findings shaped by unequal database access—this chapter argues that contemporary knowledge production is increasingly structured by geopolitical fragmentation and infrastructural asymmetries. These dynamics not only constrain collaboration but also reshape how intentions, credibility, and trust are interpreted across contexts.

In this light, the future of the BioGovernance Commons is 'pessoptimistic': marked by both the erosion of shared epistemic spaces and the renewed necessity of sustaining them. The chapter proposes 'commoning' as an ongoing, practice-based approach to maintaining dialogue under conditions of difference. Rather than assuming openness, it calls for sustained vigilance and relational work to preserve the fragile infrastructures through which collaborative understanding remains possible.

Keywords Pessopotimism · Futures · Global governance · Commoning

The BioGovernance Commons began as a modest, self-organised academic initiative in response to a moment of crisis—the rapid political and cultural transformations precipitated by the COVID-19 pandemic. It emerged as an experimental, and at times, urgent effort to revive and sustain dialogue on life science governance across the United States, the United Kingdom, and China, at a time when geopolitical tensions were already placing a strain on scientific exchange. Conceived not as a formal institution but as a relational space, the Commons sought to intervene

J. Y. Zhang (✉)
Centre for Global Science and Epistemic Justice, University of Kent, Canterbury, UK
e-mail: y.zhang-203@kent.ac.uk

J. Y. Zhang et al. (eds.), *BioGovernance Commons and China's Role in Global Risk Governance*, Emerging Globalities and Civilizational Perspectives,
https://doi.org/10.1007/978-3-032-32198-5_9

precisely where official diplomacy faltered: in the fragile yet indispensable domain of epistemic exchange. Its premise was both simple and ambitious—that even amid mistrust, censorship, and ideological divergence, sustained dialogue could help preserve the conditions for the 'good' development of the life sciences.

While the acute phase of the pandemic has receded, the sense of crisis that animated the BioGovernance Commons has not dissipated; rather, it has been reiterated and amplified through subsequent geopolitical developments. To borrow the term from China studies scholar William A. Callahan (2010), the future of the BioGovernance Commons may be understood as 'pessoptimistic'. That is, shifting configurations of material power, epistemic aspiration (i.e. the desire or commitment to know, understand, or inquire better) and political authority—exemplified by China's rise in global science and its expanding influence in the global socio-cultural sphere—have simultaneously generated both hope and anxiety. Crucially, these dynamics are not uniformly experienced: such aspirations and insecurities are differentially produced, interpreted, and evaluated across communities, sectors, and national contexts. As a consequence, the foundational premise of the Commons now appears both more precarious and more necessary than ever. Two events from early 2026, observed during the final stages of editing this volume, illustrate this precariousness and the renewed need for a commoning approach.

Horizon Europe and the Restructuring of EU–China Scientific Collaboration

The first event I want to note is that, in 2026, the European Union moved to exclude or significantly restrict Chinese entities from participating in substantial parts of Horizon Europe—the EU's €93.5 billion research and innovation programme (Fieldhouse 2026). These measures primarily target strategically sensitive areas of research and are justified on the grounds of protecting intellectual property, safeguarding research security, and reducing risks associated with dual-use technologies.

The policy trajectory leading to this shift can be traced back to the period when the BioGovernance Commons was initiated. At that time, negotiations were underway for an EU–China 'Joint Roadmap' for scientific cooperation, launched around 2019 following the fourth Innovation Cooperation Dialogue. The aim was to establish a more balanced framework to ensure reciprocity, intellectual property protection, and fair research conditions. While intended to guide future science, technology, and innovation (STI) cooperation, these negotiations have been protracted and uneven, often reflecting broader difficulties in managing bilateral research relations, with 'distrust spilling over to Europe, changing the global research cooperation landscape' (Sharma 2021). The first concrete restrictions appeared in the Horizon Europe Work Programme 2023–2024, where Chinese participation in certain 'sensitive' topics was limited (CEIAS 2025). The policy hardened in 2025, with proposed exclusions under Article 22(6) extending to most research areas, including health, civil security, digital, industry and space, as well as selected topics in climate, culture, and the bioeconomy.

Professor Li Tang, a public policy specialist at Fudan University in Shanghai, whom I have known through her participation in the BioGovernance Commons since 2024, was cited by *Nature* as expressing optimism that these new restrictions on China will not reduce collaboration between European and Chinese organisations, in part because the EU continues to promote an 'open science' approach that encourages international collaboration. However, I think there are grounds for concern: the European Union's decision to restrict Chinese participation in Horizon Europe marks a significant shift towards a more security-oriented research policy, potentially rendering 'open science' a conditional practice rather than a prevailing norm.

More importantly, there is a need to reconfigure the risk–benefit analysis of restricting international collaboration. For example, the scale of Horizon Europe's investment in AI-related research has been estimated at approximately €1.6 billion in recent work programmes (European University Association 2025), alongside European Commission ambitions to mobilise over €3 billion annually in AI investment through combined public–private initiatives. By comparison, China's annual AI investment is estimated at around $98 billion (Mak 2025), while the broader AI industry was valued at over 1.2 trillion yuan (approximately $174 billion) in 2025 (Xinhua 2026).

This disparity can be read as part of the rationale underpinning the EU's decision to exclude Chinese entities from EU-funded AI projects, particularly in light of concerns over strategic dependence and research security. However, it also calls into question the assumed deterrent and protective effects of such measures. On the deterrence side, a more difficult question for Global North actors is: what might follow if Chinese scientific communities were to lose either the incentive or the inclination to collaborate internationally?

On the protective side, it is often observed in AI communities that 'code carries culture': technical systems embed the values, assumptions, and social contexts of their creators. Yet, beyond this, the practice of code development itself can become an epistemic fortress: when AI systems are built within increasingly separated knowledge ecosystems, they are shaped by distinct standards, training data, threat models, and validation practices. Such divergence does not necessarily make systems more secure; rather, it may produce uneven forms of robustness and new vulnerabilities, particularly where mutual intelligibility and scrutiny are reduced. In this sense, exclusion may not simply mitigate risk, but could also contribute to the fragmentation of epistemic and technical infrastructures, with uncertain implications for long-term security and governance.

It is precisely under such conditions that a 'commoning' approach, as developed through the BioGovernance Commons, becomes both analytically and practically indispensable. Rather than presuming openness or closure as fixed policy positions, commoning foregrounds the ongoing work of sustaining relational infrastructures through which knowledge can circulate with accountability, reflexivity, and mutual recognition. This is not only critical for life sciences governance, or for emerging domains such as AI-driven biology, but also for science and innovation more broadly. In a context where epistemic communities risk drifting into parallel, partially incommensurable worlds, commoning offers a way to maintain channels of dialogue, preserve interpretive bridges, and enable forms of cooperation that remain attentive to difference without collapsing into fragmentation.

Horizon Europe's restrictions are not unexpected; rather, they can be understood as part of a broader trend of 'scientific decoupling' (Liu and Huang 2026) that has unfolded over the past decade. This process involves the gradual fragmentation of previously integrated global research systems, as geopolitical tensions, security concerns, and economic competition reshape the conditions under which scientific collaboration takes place. Rather than a complete severance of ties, it typically manifests through the selective restriction of partnerships, funding access, data sharing, and researcher mobility. In this context, knowledge production is increasingly treated as a matter of national and regional strategic interest, rather than as a global public good or a shared commons.

The future of the BioGovernance Commons is thus 'pessoptimistic'. On the one hand, escalating and seemingly enduring geopolitical tensions justify a degree of pessimism regarding the prospects for sustained global scientific dialogue. On the other hand, commoning offers a more generative horizon: a means of working across geopolitical impasses to reconnect fragmented scientific and social communities, and to reconstitute the conditions under which knowledge can circulate with mutual recognition and accountability.

One Database, Two Realities

The second event I want to note here is a personal one, offering further insight into how institutional rhetorics shape mindsets and lived experiences at the societal level. It concerns two members of the BioGovernance Commons: one based in Europe, the other in China. Our European colleague recently published a major piece of work, grounded in painstaking engagement with original Chinese-language sources accessed through the China National Knowledge Infrastructure (CNKI), one of the most comprehensive databases for Chinese academic literature, via their home institution in Europe.

While the publication was widely praised in Europe, particularly for its commitment to engaging Chinese sources, it also attracted sharp criticism from a Chinese colleague. The point of contention centred on the European colleague's claim that Chinese-language scholarship on a specific topic had a relatively late start and remains underdeveloped compared to other areas of the social sciences. This claim was supported by a keyword search identifying only 15 relevant articles in CNKI between 1997 and 2022.

Replicating the same search using identical keywords, timeframes, and parameters, the Chinese colleague was 'quite shocked' to identify 34 qualifying articles rather than 15. Although this discrepancy does not fundamentally overturn the broader observation that the field remains comparatively underdeveloped, it was nonetheless interpreted as evidence of careless—if not misleading—research.

In reality, both colleagues were correct. The divergence arose from differential access to CNKI: in recent years, foreign access to Chinese-language materials has become increasingly restricted by the Chinese authorities. It is widely recognised among scholars working outside China that not only has access become slower, but the scope of accessible content has also been significantly curtailed. By contrast, scholars based in China retain full access and therefore may neither encounter nor readily appreciate the constraints faced by colleagues abroad. Such asymmetries can lead to misunderstandings of one another's intentions, and in this case, resulted in the Chinese colleague interpreting the discrepancy as evidence of sloppy, if not irresponsible, research. Once the issue of differential access was explained, the colleague expressed understanding.

What was perhaps really 'shocking', however, was that while scholars based outside China (myself included) have long been aware of such disparities in data access, it was only through the close-knit and candid exchanges within the BioGovernance Commons that the scale of this divergence became fully apparent. Without such a space of trust, these differences might well have remained obscured.

More importantly, this example illustrates the subtle yet profound effects of institutional restrictions on the flow of information. It shows how infrastructural asymmetries can shape scholarly interpretation, generate doubt, and inadvertently seed division even among those committed to transnational dialogue, such as participants in the BioGovernance Commons. In this sense, censorship does not merely limit access to knowledge; it actively reconfigures epistemic trust and the conditions under which collaborative understanding can emerge.

Incidentally, this episode of misunderstanding coincided with UK Prime Minister Keir Starmer's visit to China in January 2026. During the visit, Starmer invoked President Xi Jinping's frequently cited 'blind men and an elephant' parable to suggest that much Western discourse on China remains partial, grasping only fragments rather than the whole, and rightly emphasised the need for deeper and more comprehensive engagement with Beijing (Parker and Leahy 2026). The speech, and its accompanying video (see www.youtube.com/shorts/HxB3a6BU0Xc), circulated widely across Chinese media, including WeChat, resonating strongly with a broader public sentiment in China that the country continues to be misunderstood by the outside world.

While it is true that epistemic injustice persists, particularly in relation to emerging powers such as China, the example above points to a more difficult question for Chinese institutions and stakeholders: to what extent is this 'blind men and the elephant' phenomenon also shaped by China's own regimes of information control?

This points to a second layer of the pessoptimistic future of the biogovernance commons, not merely as a specific initiative we have experimented with, but as a broader conceptual horizon. While, as academics, we may be inclined to remain optimistic about the possibilities of cross-border dialogue and exchange, such optimism cannot be taken for granted. It must be tempered by a degree of pessimism, or more precisely, a necessary vigilance, towards the forces that are not only actively constraining avenues of exchange, but also more subtly shaping our willingness to engage with one another. This further underlines our original emphasis on seeing the global life science community not as a given, but as one that requires proactive 'commoning'. Commoning does not simply refer to bottom-up or informal interventions that operate outside institutions. Rather, it is both a method and a rationale: a way of coming together grounded in deliberation, mutual respect, and an equal emphasis on listening as on 'having a voice'. It also reframes what counts as success—not the elimination of difference, but the capacity to understand, coordinate, and even benefit from differences. As a founder of the BioGovernance Commons initiative, I consider commoning can and should be incorporated into different institutions', collectives' and individuals' ways of thinking, doing and living with others (Zhang 2023).

Concluding Words

The final editing of this volume also witnessed the passing of an intellectual giant, Jürgen Habermas, in March 2026. It marks not only the loss of a towering intellectual figure, but also a symbolic rupture in the post-war imagination of rational-critical debate as the foundation of democratic life. Habermas's conception of the public sphere, as a space in which citizens engage in unrestricted, reasoned deliberation on matters of common concerns, has long underpinned normative visions of science, governance, and international cooperation. Yet, his death comes at a time when that very ideal is under strain from resurgent nationalism, polarisation, and the erosion of liberal democratic norms.

In parallel, the global landscape of science is increasingly entangled in geopolitical competition. At a time when public discourse is increasingly marked by cynicism, strategic distrust, and communicative breakdown, commoning offers a different trajectory. It reclaims the possibility of a plural, dialogical public sphere—one that does not deny conflict, but sustains the conditions under which disagreement can remain productive rather than destructive. In doing so, it also safeguards the epistemic infrastructures upon which the life sciences depend: the ability to share, critique, and co-produce knowledge across borders.

What distinguishes the BioGovernance Commons, however, is not merely its geopolitical positioning, but its methodological commitment to *commoning*. Commoning is not the management of shared resources, but the active cultivation of relational infrastructures, spaces where differences are not resolved prematurely, but rendered intelligible and workable. In other words, the whole idea of the BioGovernance Commons was not to enforce a 'consensus' or a singular, unified pathway forward across different scientific communities. Rather, a sense of solidarity was built—and could be sustained—precisely because we emphasise and value respect for, and working with, differences. The point is not to 'overcome' or 'remove' differences; rather, it is about learning to appreciate and coordinate them.

This resonates with, yet also extends beyond, Habermasian ideals: where the public sphere presupposes conditions of openness and equality, commoning begins from asymmetry, fragmentation, and partial connection.

In this sense, the future of the BioGovernance Commons lies not in institutionalisation alone, but in its capacity to operate as a distributed, adaptive practice. It is a mode of governance that works through sustained engagement across disciplinary, sectoral, and national boundaries—enabling knowledge to circulate not as a homogenised consensus, but as a mutually enriching process of translation and negotiation. Such work is necessarily slow, reflexive, and often fragile. Yet, it is precisely this fragility that renders it indispensable in a fractured world (Namdeo and Zhang 2024; Zhang 2025).

The development of the BioGovernance Commons volume thus takes place at a critical historical juncture. What began as a small, self-organised experiment now appears as a necessary prototype for future forms of global biogovernance. If the Habermasian public sphere risks fragmentation under contemporary pressures, then commoning may offer a way to reassemble it—not as a singular, universal space, but as a network of relational practices capable of sustaining inclusive deliberation and enabling more effective, ethical, and globally attuned knowledge flows. In this light, the BioGovernance Commons is no longer only an empirical case. It is a normative proposition: that in an era of fractured worlds, the future of science governance depends not on overcoming difference, but on learning how to live—and think—together through it.

References

Callahan, William A. 2010. *China: The pessoptimist nation*. Oxford: Oxford University Press.

CEIAS. 2025. The EU realpolitik turn in research cooperation with China. CEIAS Insights https://ceias.eu/the-eu-realpolitik-turn-in-research-cooperation-with-china/.

European University Association. 2025. Horizon Europe work programme 2025: New funding opportunities for universities. https://www.eua.eu/news/eua-news/horizon-europe-work-programme-2025-new-funding-opportunities-for-universities.html.

Fieldhouse, Rebecca. 2026. Why Europe barred China from flagship horizon research projects. *Nature* 650:808. https://doi.org/10.1038/d41586-026-00435-w.

Liu, Wei and Cen Huang. 2026. The dangers of science decoupling. *UNESCO*, February 11, 2026. https://www.unesco.org/en/articles/dangers-science-decoupling.

Mak, Robyn. 2025. China's slim AI budgets are self-reinforcing. *Reuters Breakingviews*, October 8, 2025. https://www.reuters.com/commentary/breakingviews/chinas-slim-ai-budgets-are-self-reinforcing-2025-10-08/.

Namdeo, S., and J. Y. Zhang. 2024. Rethinking science diplomacy and global biosecurity: Challenges, emerging practices and the way forward. *International Affairs* 100:iiae187.

Parker, George and Joe Leahy. 2026. China rolls out the red carpet for Keir Starmer, *Financial Times* 29 January. https://www.ft.com/content/0ac69fd8-e2ab-4a94-8c64-05ed3c621ab1.

Sharma, Yojana. 2021. EU engaged in roadmap talks with China on research cooperation. *University World News*, 19 November. The EU is currently engaged, the global research cooperation landscape. https://www.universityworldnews.com/post.php?story=20211119103816587.

Xinhua. 2026. China's core AI industry scale tops 1.2 trillion Yuan in 2025: Official. *The State Council of the People's Republic of China*, March 5, 2026. https://english.www.gov.cn/archive/statistics/202603/05/content_WS69a96a11c6d00ca5f9a09932.htm.

Zhang, Joy Y. 2023. Decolonising the temporal and relational assumptions in contemporary science and science policies. *Critical Policy Studies* 17:162–174. https://doi.org/10.1080/19460171.2023.2180402.

Zhang, Joy Y. 2025. *An O.D.E.SS.I. for public engagement in a new era of science. Canterbury, UK: Centre for Global Science and Epistemic Justice,* University of Kent. https://doi.org/10.5281/zenodo.1516287.

Joy Y. Zhang is Professor of Sociology with a first degree in medicine. She is also the Founding Director of the Centre for Global Science and Epistemic Justice (GSEJ) at the University of Kent. An internationally recognised expert in the transnational governance of scientific risk, she specialises in emerging life sciences that impact both human and environmental futures. Her work advances sociological thinking on risk, cosmopolitanism, decolonisation, and subaltern politics, with a comparative focus on four major scientific powers: China, India, the UK, and the US.

Zhang is especially known for her expertise in science diplomacy: how scientific collaboration, governance, and public engagement can serve as sites for international negotiation, trust-building, and ethical coordination in times of geopolitical tension. In this regard, she has been at the forefront of efforts to connect responsible innovation, global governance, and diplomatic practice in the life sciences. She has initiated several landmark programmes, including the BioGovernance Commons (2021), the founding of GSEJ (2022)—whose research has informed G20 and G7 policy discussions—and, most recently, O.D.E.SS.I., a global 'odyssey' for public engagement with the life sciences. She is the author of three academic books and publishes widely in high-impact journals across both the natural and social sciences. She is frequently interviewed by global media and advises leading research and policy institutions in Europe and Asia.

Chapter 10
Coda I: Bridging Divides, Forging Trust: A Plant Scientist's View on BioGovernance Commons

Wei He

I am a plant biologist with education and professional experience in both China and the UK. I also serve as a champion of international collaboration at Northwest University, a comprehensive public university in Xi'an, Shaanxi Province. Founded in 1902, it is one of the oldest higher education institutions in western China and is known for its strong traditions in geology, archaeology, chemistry, life sciences, and environmental science. The university is part of China's national 'Double First-Class' initiative, with selected disciplines recognised for world-class development.

While I was not formally part of the BioGovernance Commons discussions, which mainly focused on biomedical sciences, I contributed to a series of UK–China dialogues on public engagement in the life sciences organised by Joy Zhang, Ruipeng Lei, and colleagues during 2017–2019 (Zhang 2018). I was later informed that my contributions, as a natural scientist, helped shape the early conversations that led to the founding of the BioGovernance Commons.

Although I was not directly involved in the initiative, I am pleased to read this collection, *Building a BioGovernance Commons: Engaging with China in Global Risk Governance*. In an era where geopolitical tensions increasingly cast shadows over scientific collaboration, the book's commitment to sustaining open exchange—and to grappling with its challenges—is both timely and important. It moves beyond polarised narratives that often frame China–West scientific relations, offering instead a nuanced and empirically grounded account of how dialogue can bridge cultural, institutional, and political divides. This spirit of constructive engagement strongly resonates with the exchanges I experienced during the 2017–2019 discussions. As a biologist with a deep appreciation for cross-cultural collaboration, I find the volume not only intellectually rigorous but also a compelling call to re-centre

W. He (✉)
Office of International Cooperation, Northwest University, Xi'an, China
e-mail: weihebio@nwu.edu.cn

J. Y. Zhang et al. (eds.), *BioGovernance Commons and China's Role in Global Risk Governance*, Emerging Globalities and Civilizational Perspectives,
https://doi.org/10.1007/978-3-032-32198-5_10

science on its core mission: advancing human welfare through cooperation rather than division.

As a scientist and as an Executive Deputy Director for a university's international outreach, I know only too well how simplistic binaries—'authoritarian China' versus 'liberal West', national sovereignty versus global solidarity—hurt research, for both China and the world (Martins and Schwaag Serger 2023; Shih and Forsberg 2023). From high-profile research misconduct cases to tensions between top-down regulatory frameworks and bottom-up scientific practice, I appreciate the fact that the chapters collectively offer a critical yet empirically grounded account of China's emergence as a global bioscience powerhouse. Chapters like Lu Gao's exploration of COVID-19 research in China, for instance, humanise the 'nationalist science' narrative by highlighting the quiet heroism of researchers navigating international scepticism, domestic censorship, and public misunderstanding—many of whom collaborated across borders to develop vaccines and antibodies (Chap. 2). Similarly, Haidan Chen's analysis of Chinese biobanks reveals a system grappling with dual pressures: safeguarding genetic sovereignty against historical fears of biopiracy, while striving to integrate with global research networks (Chap. 4). These accounts remind us that behind every 'Chinese science' headline are individual scientists, ethics committees, and policymakers grappling with the same dilemmas as their Western counterparts.

The book's conceptual anchor, the 'BioGovernance Commons', is both thoughtful and generative. As the editors noted in Chap. 1, the concept is developed from Elinor Ostrom's work on commons as dynamic, relational practices rather than static resources, framing transnational scientific collaboration as an ongoing process of 'commoning': negotiating differences, addressing power asymmetries, and building trust through sustained engagement (see Chap. 1).

At the same time, all four editors, the founders of the initiative, are social scientists. As a scientist, for me, the notion of 'BioGovernance Commoning' evokes an additional layer of meaning. Research trajectories, particularly in cutting-edge discovery science, cannot be straightforwardly planned and often resist top-down direction. Breakthroughs more often emerge through what might be called *scientific commoning*: collective processes of exchange, negotiation, and iterative revision of aims, methods, and tools.

Science, in a sense, is not 'universal'. Its development cannot be standardised or fully predicted; it depends on spaces that enable open-ended deliberation and intellectual risk-taking. The BioGovernance Commons initiative offers a compelling illustration of how such spaces might be cultivated in practice, especially at a time when global dialogue is narrowing, scepticism is deepening, and science is increasingly subject to national securitisation. The BioGovernance Commons' engagement with the aftermath of He Jiankui's controversial work, for example, shows that when scientists, ethicists, and policymakers convene in settings that prioritise dialogue over political positioning, it becomes possible to develop responses that are both context-sensitive and collectively meaningful (Cyranoski 2019; Zhang et al. 2023).

As someone who has worked extensively across cross-cultural scientific collaborations, I am keen to underline a long-standing but often overlooked point—though one that is easier said than done: science is most generative when it is not overdetermined by political posturing. Yet, contemporary global bioscience is frequently entangled in geopolitical anxieties, manifested in restrictive data-sharing regimes, heightened suspicion of 'foreign' research agendas, and the instrumentalisation of scientific misconduct as a proxy for broader political contestation (Kaye et al. 2009; Phillips et al. 2020). The BioGovernance Commons initiative offers a careful but important counterpoint, demonstrating how sustained, equitable dialogue can create space for collaboration that is neither naïvely apolitical nor unduly constrained by geopolitical divisions. In this regard, I was particularly drawn to the chapter by Li Du and Zhangyu Wang (Chap. 5) on cross-border genetic data transfer in China. Their analysis reveals the frictions produced by overlapping and, at times, internally inconsistent regulatory frameworks, frictions that can impede collaboration without necessarily strengthening oversight, while also pointing towards more coherent ways of reconciling national sovereignty with global scientific interdependence. Similarly, Simona Novaretti's comparative legal analysis challenges the enduring trope of a 'China without law', foregrounding instead how provisions within China's Civil Code, particularly those relating to human dignity, can contribute meaningfully to global biogovernance debates when engaged with on equal terms (Chap. 6).

One gap in the BioGovernance Commons initiative is its engagement with industry. To be sure, the role of non-public funding in China's development of biotechnology is relatively recent, but it has grown rapidly, especially over the past 5 years (Xiao and Kerr 2022; Hess 2025). While the current focus on academia, policy, and science communication is already impressive, I look forward to the BioGovernance Commons continuing to grow and expanding further to include more actors from the bio-industry, both within and beyond China. Alternatively, this gap may serve as an incentive to inspire more BioGovernance Commons–like initiatives to emerge.

One key takeaway from this volume is that scientific governance cannot be a one-size-fits-all enterprise, nor can it be dictated by a single set of cultural or political norms. The book argues compellingly that the future of global bioscience depends on creating spaces where non-Western perspectives are not just included but centred: where China's emphasis on relational ethics, India's approach to biodiplomacy, and Nigeria's grassroots biosecurity practices are recognised as valuable contributions to a shared governance ecosystem (Chap. 8). This is not about abandoning principles of transparency or accountability; it is about recognising that these principles can be operationalised in diverse ways that reflect local contexts.

In conclusion, *Building a BioGovernance Commons* is more than an academic text—it is a manifesto for a more inclusive, collaborative future for science. It challenges scientists, policymakers, and funders to reject the zero-sum logic of geopolitical competition and invest in the slow, painstaking work of dialogue. For researchers like myself, it offers a roadmap for navigating cross-cultural collaborations with humility and curiosity, rather than suspicion. In a world where pandemics, bioterrorism, and technological disruption require global solutions, this book is essential reading—reminding us that science's greatest strength lies not in its ability to produce universal truths, but in its capacity to bring people together in pursuit of them.

For successful science governance, it is worth recalling John F. Kennedy's injunction in his inaugural address on 20 January 1961.: 'Let us never negotiate out of fear. But let us never fear to negotiate.' Reframed for today's context, the task is not to 'oppose any foe' in a geopolitical sense, but to confront the more diffuse challenges facing global science—polarisation and binary thinking, opacity and lack of accountability, and the increasing geopolitical instrumentalisation of research. Against this backdrop, the 'commoning' practices documented here offer a vital alternative. They suggest that when dialogue is prioritised over positioning, and relational work over rivalry, science can once again function as a bridge rather than a battleground in an increasingly divided world.

References

Cyranoski, D. 2019. What CRISPR-baby prison sentences mean for research. *Nature* 577:154–155.

Hess, Sean. 2025. *Year of the NewCo? China's emerging impact on global drug development.* Bristows. https://inquisitiveminds.bristows.com/post/102k2zc/year-of-the-newco-chinas-emerging-impact-on-global-drug-development.

Kaye, Jane, Catherine Heeney, Naomi Hawkins, Jantina de Vries, and Paula Boddington. 2009. Data sharing in genomics—Re-shaping scientific practice. *Nature Reviews Genetics* 10 (5): 331–335. https://doi.org/10.1038/nrg2573.

Kennedy, J. F. 1961. *Inaugural address*, 20 January. https://www.jfklibrary.org/learn/about-jfk/historic-speeches/inaugural-address.

Martins, Igor, and Sylvia Schwaag Serger. 2023. An age of disentangled research? *Issues in Science and Technology* 40 (1): 38–43. https://doi.org/10.58875/WBCQ3772.

Phillips, Mark, Fruzsina Molnár-Gábor, Jan O. Korbel, Adrian Thorogood, Yann Joly, Don Chalmers, David Townend, and Bartha M. Knoppers. 2020. Genomics: Data sharing needs an international code of conduct. *Nature* 578 (7793): 31–33. https://doi.org/10.1038/d41586-020-00082-9.

Shih, T., and E. Forsberg. 2023. Origins, motives, and challenges in Western–Chinese research collaborations amid recent geopolitical tensions. *Higher Education* 85:651–667.

Xiao, Zhenzhen, and William A. Kerr. 2022. Biotechnology in China—Regulation, investment, and delayed commercialization. *GM Crops & Food* 13 (1): 86–96. https://doi.org/10.1080/21645698.2022.2068336.

Zhang, Joy Y. 2018. *Governing scientific accountability in China. Final report of the ESRC research project.* Canterbury: GSA-China. https://kar.kent.ac.uk/77924/1/GSA-Project-Report.pdf.

Zhang, Joy Y, Ruipeng Lei, Kathleen Vogel, and Sonia Ben Ouagrham-Gormley. 2023. *Looking back into the future: CRISPR and social values: BioGovernance common's open discussion with Chinese academics.* 13 February. Canterbury: Centre for Global Science and Epistemic Justice, University of Kent.

Wei He is an Associate Professor and Executive Deputy Director (Associate Dean) of the Office of International Cooperation (School of International Education) at Northwest University, where he also serves as a doctoral supervisor. He is recognised as a 'Western Light.' Young Scholar by the Chinese Academy of Sciences.

Chapter 11
Coda II: Scaling the Commons: The Future of Shared Biogovernance from the Perspective of a Synthetic Biologist

Xian-En Zhang

At the editors' kind invitation to review and contribute a commentary to this volume, I was initially hesitant. As a biotechnologist, I am not accustomed to the discourse of the humanities and social sciences. Yet, encouraged by Joy to share insights from my own research, I gladly accepted. Having read the volume in full, I find it clear that, against the backdrop of deglobalization and increasingly complex geopolitics, its authors have adopted a broad vision and forward-looking perspective, actively contributing to the construction of the 'BioGovernance Commons.' I deeply appreciate this far-sighted approach. Inspired by this, I would like to offer my humble observations drawing on my experience in advancing synthetic biology, with a focus on global biogovernance.

Understanding of BioGovernance Based on Practices in Synthetic Biology

To help readers orient where my observations come from, it may be useful to first say a bit about my field of expertise, synthetic biology, and briefly outline my experience in the transnational discussion and governance of this field.

Since the advent of genetic engineering, bioethics, biosafety and biosecurity have become major issues of global concern. With the efforts of the international

X.-E. Zhang (✉)
Faculty of Synthetic Biology, Shenzhen University of Advanced Technology, Shenzhen, China

Institute of Biophysics, Chinese Academy of Sciences, Beijing, China
e-mail: zhangxe@ibp.ac.cn

J. Y. Zhang et al. (eds.), *BioGovernance Commons and China's Role in Global Risk Governance*, Emerging Globalities and Civilizational Perspectives,
https://doi.org/10.1007/978-3-032-32198-5_11

scientific community, countries have gradually established regulatory systems suited to their national contexts.

Synthetic biology is an emerging biotechnology whose enabling technologies are rooted in genetic engineering and systems biology. They encompass a suite of advanced tools, including genetic circuit construction, codon expansion, precision gene editing, genome design and synthesis, AI-assisted de novo protein design, and chassis cell optimisation. These developments have propelled the biological sciences from understanding life toward a new stage of artificially designing and synthesising life. The field pioneers the transformative build-to-learn paradigm, reshaping the fundamental biological research model. It also establishes the landmark build-to-use framework, underpinned by iteratively upgraded enabling technologies. Such advances unlock far-reaching applications beyond the scope of conventional biotechnologies. However, as a transformative technology, synthetic biology poses new challenges to traditional boundaries in bioethics, biosafety governance, and ecological risk control. If GMOs refer to genetically modified organisms, synthetic biology holds the potential to create novel biological forms, thus giving rise to broader ethical and safety concerns.

In global governance of emerging technologies, China has gradually shifted from a passive participant to a co-builder and leader, accumulating systematic experience in fostering international consensus and advancing domestic governance practice.

From 2011 to 2012, the Royal Society and Royal Academy of Engineering of the United Kingdom, together with the National Academies of Sciences and Engineering of China and the United States, held a series of symposiums on synthetic biology in London, Shanghai and Washington, respectively. These events provided a high-level global platform for academic and policy exchange, focusing on enabling technology innovation, social empowerment, and governance principles in synthetic biology. The U.S. National Academies later released the six-party symposium summary report entitled *Positioning Synthetic Biology to Meet the Needs of the twenty-first Century*, which included systematic presentations of the views and contributions of Chinese scholars, as well as my presentation on China's synthetic biology roadmap.

In 2014, entrusted by the Chinese Academy of Sciences, I joined the Working Group on Synthetic Biology of the InterAcademy Partnership (IAP), the global network of science academies. The working group comprised six experts from the Royal Society (RS), the National Academy of Sciences (NAS), the Chinese Academy of Sciences, the European Academies' Science Advisory Council (EASAC), the Science Council of Japan (SCJ), and the Global Young Academy (GYA). Its key mission was to address international concerns regarding the development of synthetic biology. Following extensive discussions, the IAP issued a formal statement affirming the significant application potential of synthetic biology in healthcare, food security, energy supply, and other fields. It called for the coordination of scientific and technological innovation with biosafety and biosecurity, advocated a science-based governance model, emphasised the strengthening of research ethics and public engagement, promoted global scientific cooperation and regulatory harmonisation, and strived to achieve responsible and collaborative innovation.

In 2018, China launched the National Key Research Program on Synthetic Biology, with dedicated arrangements for research related to ethics and safety. Over nearly a decade of implementation, beyond a string of major scientific breakthroughs, the program has delivered important outcomes in ethical assessment, risk prevention and control, science popularisation, and talent training. Many authors of this book have been deeply involved in these efforts.

In 2024, the *Synthetic Biology Roadmap 2030* was developed and finalized under the coordination of the author and over 120 top experts. This roadmap contains a special chapter on governance principles, defines regulatory redlines and implementation standards, and establishes a full-chain governance system. The English version of this roadmap was published in open access (OA) in March 2026, offering the world a feasible and practice-oriented Chinese approach.

In addition, I have long delivered popular science lectures on synthetic biology to middle and high school students, university students, entrepreneurs, and government administrators. While different audiences focus on distinct topics, bioethics and biosafety have consistently remained core issues of shared concern.

China's Pursuit of Scientific Progress Merits a Fair and Inclusive Environment

I am frequently approached by international peers in natural and social sciences, alongside policymakers and research funders, asking how to interpret China's role in global science. Observers still need time to fully grasp the development trajectories of late-developing new reearch participants. At the same time, as with any country, China's pursuit of scientific progress warrants fair and informed discussion.

Since the reform and opening-up, China's scientific research capacity has continuously improved, achieving a qualitative leap over the past two decades. In terms of academic achievements, from 2024 to 2025, China published a large number of papers indexed in Web of Science annually, with the number of high-impact papers ranking first globally. In the Nature Index, China has remained the world's top since 2022, performing prominently in chemistry, material sciences and other fields, while biological sciences and health sciences have risen to the second and third places globally, respectively. In synthetic biology (sometimes referred to as engineering biology), China has led the world in the total number of papers and highly cited papers since 2022, based on data retrieved from Web of Science (Clarivate) on April 11, 2026. Chinese researchers, from a non-English-speaking country, face notable linguistic barriers, obstacles that Western scholars rarely perceive.

Significant progress has also been made in bioscientific data. In 2021, China's annual deposition of protein structures in the Protein Data Bank (PDB) ranked second worldwide (Wu 2025), and since 2022, China's Protein Data Bank (PDB China) has formally joined the Worldwide Protein Data Bank (wwPDB) as its fourth member from a national or regional entity (Ji et al. 2025). Meanwhile, China has ranked among the top two countries worldwide in terms of global genomic base data submitted to the International Nucleotide Sequence Database Collaboration (INSDC) (Convention on Biological Diversity 2020), and the National Genomics Data Center of the China National Center for Bioinformation (CNCB-NGDC) aligns its data standards with the INSDC and regularly mirrows the complete set of public publicly available INSDC sequence records (CNCB-NGDC 2026).

China's rapid scientific and technological development has profoundly reshaped the global innovation landscape, while also eliciting diverse perceptions and responses from traditional holders of technological discourse power. On the whole, such responses have been largely positive. Regrettably, however, some negative reactions have steered this process toward geopolitical competition and national security issues, leading to the imposition of technological barriers and restrictions on biological data sharing and academic exchanges, thus exerting a significant impact on the global ecosystem of scientific collaboration. For instance, the number of jointly authored papers between China and the United States in biological sciences and biotechnology has fallen by 35% over the 5 years since 2020, according to data accessed from Web of Science (Clarivate) on April 11, 2026.

These politicised and discriminatory measures clearly lack legitimate justification. In fact, without China's in-depth participation and practical experience, global bio-governance can hardly establish a truly comprehensive, balanced, pragmatic and feasible governance system.

Constructing the BioGovernance Commons Needs China's Contribution

In the volume, I'm delighted to see that the authors address the deep-seated contradictions in global biogovernance and propose the concept of the 'BioGovernance Commons', based on a multidimensional analysis.

First, focusing on core concerns in global biogovernance, this volume examines major concerns including public health emergencies, biological data flow and genetic resource sharing. It also accommodates the needs of TCM internationalisation and seeks to align a universal governance framework with local practices.

Second, amid structural challenges such as the inadequate implementation of the One Health approach, it reveals the constraints of existing international mechanisms and profound global divergences of interest. The EU adopts process-based regulation and ex-ante risk control over GMOs, while the United States follows a product-oriented model with outcome-based standards. As for China, it retains its fundamental legal system governing GMOs; meanwhile, the country has actively revised its regulations on genetically modified crops in recent years and introduced a tiered regulatory framework aligned with the international Site-Directed Nuclease

(SDN) classification of gene-edited products to balance biosafety and innovation. Environmental stakeholders emphasise rigorous risk prevention, whereas research and industry sectors endorse evidence-based governance (Tachikawa and Matsuo 2023). The international community has reached consensus on the responsible advancement of human somatic genome editing therapy, the strict prohibition of human heritable genome editing (WHO 2021), and the prevention of technological misuse. Screening for dual-use technologies, oversight of high-risk sequences, and mutual recognition of global standards remain central to future cooperation.

Third, in response to biased perceptions concerning China's biogovernance philosophy, it employs empirical evidence to clarify misunderstandings and challenge prejudice.

Fourth, from the standpoint of emerging economies, it elevates theoretical discourse through original achievements in national legal and ethical studies, fosters integration and mutual learning between these economies and established governance actors, and expands the intellectual and diplomatic space for global biogovernance.

Fifth, by transcending the rigid game logic inherent in traditional governance arrangements, it establishes an inclusive mechanism underpinned by sustained dialogue, relational norms, and mutual trust and accountability, providing a pragmatic and flexible pathway to bridge cognitive divergences and dismantle barriers to cooperation.

To sum up, this volume systematically elaborates the logic for building a biogovernance commons, analyses the opportunities and challenges facing global research governance, and highlights the theoretical innovation and sense of responsibility of emerging economies, including China. It provides important support for the establishment of an open, inclusive, equitable, and barrier-free global biogovernance system.

In light of the above, I also put forward two suggestions for future BioGovernance Commons activities or similar transnational endeavours to consider:

(1) As exemplified by the series of six-academy symposia held in 2011 and 2012 (see https://royalsociety.org/news-resources/projects/synthetic-biology/), there are tremendous benefits to both the science and its governance from holding regular high-level forums, refining governance frameworks design, and developing clear implementation roadmaps, thereby enhancing international influence of global scientific governance.
(2) There is also a need to expand the scope of social and regulatory discussions, and to draw connections where appropriate, in line with the growing breadth and intersection of emerging scientific fields. For example, future BioGovernance Commons discussions and similar initiatives could incorporate cutting-edge technologies, including synthetic biology, stem cell therapy, therapeutic gut microbiome and microphysiological systems into the governance system, and promote synchronised adaptation of governance architecture with technological development.

Conclusion

I echo Joy's point that the model of joint contribution and shared governance in the biological field is both fragile and indispensable. This volume provides new perspectives for reshaping the global biogovernance system amid a fragmented landscape. In the long run, the sound development of global biogovernance requires an international environment that is open, inclusive, culturally diverse, equitable, and jointly governed. Only by safeguarding the bottom line of innovation through professional governance and empowering biological sciences through openness and cooperation can we ensure that cutting-edge technologies benefit all humanity.

In China, innovation-driven development has become a national strategy. R&D investment has sustained rapid growth, and the headcount of full-time equivalent (FTE) R&D personnel has surged, hitting more than seven million in 2024 (*National Bureau of Statistics of China* 2025), a figure far exceeding that of any other economy, with its scientific outputs ranking among the world's top tier. With this momentum continuing to grow, China will play an increasingly important role within the framework of the inclusive, multicultural BioGovernance Commons advocated in this volume, steering global governance from 'pessoptimism' toward a bright future.

References

CNCB-NGDC. 2026. Database resources of the national genomics data center, China national center for bioinformation in 2026. *Nucleic Acids Research* 54 (D1): D28–D47.

Convention on Biological Diversity. 2020. *Combined study on digital sequence information in public and private databases and traceability (CBD/DSI/AHTEG/2020/1/4)*. Montreal: Secretariat of the Convention on Biological Diversity.

Ji, Wenxin, et al. 2025. Protein data bank China: Grounding in the present, focusing on the future. *Chinese Bulletin of Life Science* 37 (7): 887–893.

National Bureau of Statistics of China. 2025. *China Statistical Yearbook*.

Tachikawa, M., and M. Matsuo. 2023. Divergence and convergence in international regulatory policies regarding genome-edited food: How to find a middle ground. *Frontiers in Plant Science* 14:1105426. https://doi.org/10.3389/fpls.2023.1105426.

WHO Expert Advisory Committee on Developing Global Standards for Governance and Oversight of Human Genome Editing. 2021. *Human genome editing: A framework for governance*. Geneva: World Health Organization. Licence: CC BY-NC-SA 3.0 IGO.

Wu, Jiarui. 2025. Ten-year for national facility of protein science in Shanghai. *Chinese Bulletin of Life Science* 37 (7): 748–754.

Xian-En Zhang is Dean and Chair Professor of the Faculty of Synthetic Biology at Shenzhen University of Advanced Technology, and a retired Professor at the Institute of Biophysics, Chinese Academy of Sciences. His interdisciplinary research interests span synthetic biology, biosensors and nanobiology, addressing basic problems in virology, cancer biology and cell biology, with over 400 publications. He received his B.Sc. from Hubei University (1982), and later earned his M.Sc.

in Microbiology and Ph.D. in Biochemistry from the Chinese Academy of Sciences, as well as an Honorary Doctor of Science Degree from the University of Alberta (2015).

He is a Fellow of AIMBE and RSC. He serves as Vice President of the Chinese Society of Biotechnology and Chair of its Synthetic Biology Division, Founding Co-Chair of the AFOB Division of Nanobiotechnology/Biosensors/Biochips, and Founding Chair of the ACS Guangdong Chapter, and has been a committee member of the Biosensors Congress since 2008. He led the formulation of *Synthetic Biology Roadmap 2030—Core Driving Force for Biomanufacturing* (Science Press 2026, open access).

Chapter 12
APPENDIX 1: The Ethics of Protecting 'CRISPR Babies': An International Roundtable Summary Report (2022)

Joy Y. Zhang

Introduction

The international roundtable was convened as a response to two leading bioethicists, Ruipeng Lei and Renzong Qiu, who proposed a 'special protection' to be set up in China for the world's first three genome-edited children,' a new group of vulnerable population' (Mallapaty 2022). Describing the three children, Lulu, Nana and Amy, as the result of Jiankui He's 'genetic mayhem', Lei and Qiu proposed that a proactive protection plan was warranted by 'moral obligations for future generations'.

The Centre for Global Science and Epistemic Justice (GSEJ) invited 10 panellists from the life sciences, bioethics, anthropology, sociology and patient and stakeholders' groups to reflect on what constitutes the ethics of protecting genome-edited children and (future) individuals in comparable situations, and on how open and inclusive deliberation on this issue can take place in and with China. The event was held on 18th March 2022 on Zoom. It was chaired by GSEJ's Director, Dr. Joy

Editorial note: This piece forms part of two major international discussions engaging with perceived divergences between Chinese actors and global counterparts on the governance of human genome editing. Convened through the BioGovernance Commons initiative, these dialogues did not aim to force consensus, but rather to enhance transparency, surface underlying logics and assumptions, and foster mutual understanding. In doing so, they sought to collectively identify more sensitive and responsible pathways forward in addressing this ethically and politically complex issue. This report was originally published online on 5 April 2022 by the Centre for Global Science and Epistemic Justice at the University of Kent. The views expressed in this event are the independent views of the participants and do not reflect the views of any institution.

J. Y. Zhang (✉)
Centre for Global Science and Epistemic Justice, University of Kent, Canterbury, UK
e-mail: y.zhang-203@kent.ac.uk

J. Y. Zhang et al. (eds.), *BioGovernance Commons and China's Role in Global Risk Governance*, Emerging Globalities and Civilizational Perspectives,
https://doi.org/10.1007/978-3-032-32198-5_12

Zhang, who had a series of in-depth conversations with academics in China in the 5 weeks leading up to the roundtable to understand the evolving situation. Six other participants who were involved in these conversations were also invited to attend the meeting.

We chose the format of a roundtable deliberately so as to give equal importance to all perspectives. The roundtable proceeded in alphabetical order with one exception. In recognition of the importance of patients' perspectives, we invited Jay Johnson, who experienced long-term medical follow-up after gene therapy, to open up our discussion.

Panellist Input

Jay Johnson, Director of Volunteers at Action Wellness, participated in a small study of the Sangamo BioSciences therapy in 2010 (Tebas et al. 2014). The study enrolled 12 people infected with HIV in an open-label, nonrandomised first study of 'gene editing' in humans. Scientists removed cells from the patients using a zinc-finger nuclease (ZFN) to permanently disable these cells' CCR5 gene and then dripped the cells back into their bloodstreams. To become HIV free was what motivated Johnson to participate in this clinical study. He was the first participant who had an adverse reaction after receiving the edited cells. This included chills and stomach pain. However, reflecting on his experience, Johnson highlighted the care and responsiveness of medical staff, and commented that 'the unknown is scary but [the staff at the University of Pennsylvania, where the trial took place] made it so much easier for me to go through the whole process'. Following the treatment, Johnson had 10 years of medical follow-up at the University of Pennsylvania. He described the follow-up appointments as 'very short and sweet', which allowed him to carry on working and enjoying daily life as normal. He also expressed strong support for similar clinical studies on diseases which have a large impact on society.

Sarah Chan, a bioethicist at the Usher Institute, University of Edinburgh, commented on both the procedural and substantive aspects of protecting CRISPR babies. She considered it commendable that Chinese bioethicists were proactively taking steps to try to address this issue. However, in its current form, the call for protection risked perpetrating exactly the type of discrimination that they sought to prevent. Firstly, the proposal's framing of protections as justifiable on the basis of the well-being of hypothetical future people could be dangerous, as it diverts the focus away from the people who were in need of protection today and could, conceivably, end up harming the three children in China. Secondly, 'genomic over-protectionism' needed to be avoided. This is to say, bioethicists or experts in any single area should not arrogate to themselves the responsibility and ability to make decisions about what should or should not happen in people's lives. Rather, public dialogue is needed. We need not have discussions about what happens to people without those very people being part of the discussion. To avoid genomic over-protectionism is also to avoid possible coerced eugenics. Babies born naturally may also have different types of mutations, but are not subjected to surveillance. So, what justifies putting individuals born as a result of genome editing under increased reproductive surveillance, and for whose benefit is this? Lei and Qiu's proposed protections are unconvincing on these questions. Vulnerability itself can be a 'vulnerat-ing' concept which denies individual agency. Thirdly, there should be a clarification about whether we are protecting people from being subject to improper human agency or whether we are protecting people from genetic risks. Chan argued that we should be focusing our efforts on current and near-future people and that we should be protecting them from the type of bio-surveillance that was suggested by the draft proposal.

Ryan Ferrell joined Jiankui He's team as a consultant in August 2018. As documented in the book, *The Mutant Project* (Kirksey 2020), Ferrell worked to slow down He's research, but was only able to stop one implantation attempt. At the roundtable, Ferrell shared valuable insights on the concerns of the families involved in He's study. Ferrell first highlighted that it is very easy to downplay the stigma and daily plight that the HIV communities in China are facing, including job insecurity and restricted access to ordinary medical care. This contributes to a sense of abandonment by Chinese and Western societies, as reflected in three participants' letters to a Chinese court in February 2019. Ferrell suggested these families want access to a compassionate and independent healthcare team, which would provide basic and speciality health care, such as genetic counselling, as and when relevant. The two families have reason to fear simple primary care visits in the context of HIV and an unsympathetic global response. Ferrell also pointed out that a moratorium on the technology evades difficult discussions for the broader field. One key lesson from Jiankui He's case was that top-down governance, both in China and in Western scientific bodies, faltered and has not self-corrected. Ferrell called for a reflection on the human factors in research culture that affected this incident.

Benjamin Hurlbut, Associate Professor in the School of Life Sciences at Arizona State University, first echoed Chan's view that no responses, however well-intended and well-designed, ought not be imposed on children in the name of responsibility. He contended that just as these children were once a side demonstration for a technical project, they also risk becoming a side demonstration for ethical responsibility. He warned about the possibility of these children becoming a platform for constructing 'regimes of response', either in the name of atonement or of rectification, which turns them into an instrument in the service of bioethics rather than the other way around. Hurlbut thus went further and challenged the notion of 'we' in experts' framing of the responsibility to proactively intervene. He argued that the most fundamental condition to assist these families was to recognise that one can only become part of 'we' in sharing responsibilities with these families at the families' invitation. Hurlbut proposed that the kind of healing that this situation demanded was a 'responsibility to understand'. What is needed is not fetishising the difference of how these children were born, but a kind of humility to turn the gaze back upon ourselves to understand how this has happened. Regarding an elite-led protection proposal, Hurlbut cautioned that the imagination of a heroic caring and saving was what created CRISPR babies in the first place. Hurlbut instead called for contextualised understanding and serious interrogation of what the pathology was, for which a risky intervention was considered a compelling treatment. This is a reflection needed globally.

Hubert Kang, who oversees the operation of Institutional Review Boards of BGI-Shenzhen and of China National GeneBank, provided updates on China's regulatory development since *the Second International Summit on Human Genome Editing* (https://www.nationalacademies.org/projects/PGA-STL-17-03) in December 2018. China has approved gene therapy clinical trials for treating lung cancer, head and neck cancer, melanoma, lymphoma, HPV, and HIV. The latest development was China's 2021 approval of EdiGene's CRISPR/Cas 9 gene-editing hematopoietic stem cell therapy for sickle-cell disease (Business Wire 2021). The Ministry of Science and Technology, the National Health Commission and the State Administration for Market Regulations are key national regulators for genome editing in China. One month after Jiankui He announced his experiment, China's National Natural Science Foundation expanded its bioethical requirements for all applications. A few months later, China's National Science and Technology Ethics Committee (Jia 2019) was established. The city of Shenzhen, where Jiankui He was based, further strengthened its local review system by promulgating city-wide *Ethical Review Guidelines of Biomedical Research Involving Human Participants* (Shenzhen Municipal 2020). In 2020, China revised its *Civil Code* and criminal law (Cyranoski 2019; Song and Joly 2021), which recognises unethical experiments on genes in adults or embryos as a violation of a person's fundamental rights. China's first national biosecurity law (Lew 2021) also banned the unethical collection, preservation, use and outbound supply of genetic resources.

Eben Kirksey, an anthropologist at Alfred Deakin Institute in Melbourne, first pointed to the long tradition of thinking about medical morality in China and rebuked a Western-centric view on bioethics. He recounted how his extensive fieldwork in China helped him understand the values that shaped Jiankui He's experiment. He argued that the public values and discourse that have shaped the 'China Dream' were also reflected in the ethical conversation about disruptive innovation and the entrepreneurial spirit Jiankui He embraced. For Kirksey, both the making of CRISPR babies and the protection for them are not just about an autonomous consenting subject. It's very important to expand the scope of ethical investigation to take into account how medical morality is enmeshed in a network of human relationships. For example, the real-life cost of social stigma for the HIV community in China (such as employment discrimination) should be taken into account. Kirksey also drew the comparison between Louise Brown (the first human born after in-vitro fertilisation), who was a public figure the moment she was born and the three CRISPR babies whose identities were, as an important ethical decision made within the lab, well-protected from the start. Similar to other panellists, Kirksey opposed too much protection or surveillance. He applauded Jiankui He for offering the children insurance plans for the first 18 years of life, but noted that the insurance companies would not issue the plans because the babies were born prematurely. Kirksey encouraged further discussions about how funds from He might be used to reimburse the ongoing health care expenses for the three children. He advocated evidence-based discussions, where potential health issues can be identified by and responded to by limited new genomic studies. He considered that Chinese bioethicists had a leadership role in making consequential recommendations about this particular case and about this field broadly. He maintained that key values of China's long tradition of medical morality, which may differ from Christian-based bioethics, are worth highlighting when recommendations are made at the national and international levels.

Ruipeng Lei, the lead Chinese bioethicist who championed the special protection plan for the three genome-edited children, elucidated some of the background of the draft proposal as well as recent public discussions held in China. Lei stated that the moral imperative she and Qiu felt in drafting a protection plan was rooted in the fact that the health risks faced by the three children were essentially 'unknown unknowns'. It's an unprecedented uncertainty for the world. She emphasised that the draft recommendations were only 'a preliminary attempt to start a sustainable future conversation' and that it was 'not intended to be a concluding piece'. The discussion within the Chinese bioethics community started on 8 January 2022 at Renmin University when a small group of social scientists joined a bioethics forum on how to protect genome-edited children. On 12 January, Lei made a short presentation at a Sino-UK symposium on the legal aspect of genome editing, which was sponsored by the UK Embassy in Beijing and in collaboration with the Nuffield Council. On 13 March, Lei's research centre at Huazhong University of Science and Technology and the Centre for Bioethics at Xiamen University co-hosted a webinar on how to treat germline edited persons. The event was free and open to the public. Nearly 140 participants with different disciplinary backgrounds attended. A key theme of the 13 March webinar was how 'the best interests' of the children should be identified. Lei confirmed that most of her colleagues in China were 'strongly in support of further research and consultation on this issue'. Roundtable participants, Di Zhang, Haidan Chen, and Hui Kang, were particularly keen to avoid a one-sided view, driven either by scientific or bioethical experts. Lei also stressed that the views of the family and of the three children should always be given due consideration.

While Lei recognised Sarah Chan's point that bioethical discussions needed to keep an appropriate balance between the needs of current and future generations, Lei insisted that uncertainties posed by germline editing warranted extra precaution. The damage to genomes caused by editing could be severe but hard to detect, identify and measure with current technology. Given that public discussion has just been initiated in China, Lei considered it important to ask questions rather than provide answers:

- How does germline genome editing affect our gene pool and the well-being of future generations?
- How do we evaluate germline genome editing from different moral and cultural perspectives?
- What moral duties do the present generation have for the future generation?
- What kinds of genetic mayhem will be caused by the heritable CCR5 genome editing?

- Are the three genome-edited children entitled to be classified as a vulnerable group?
- What are the benefits or harms of being classified as a vulnerable group?
- How do we identify the 'best interests' of the three genome-edited children? Who should make these decisions?
- How do we define special protection? What is the moral distinction between protection and restriction? How do we observe this distinction in practice?
- What are the specific mental and social risks for genome-edited children?
- What kind of special protection is morally justifiable and practically feasible?
- Who should be responsible for offering special protection?
- Is it necessary for us to establish a mechanism to care for them?
- Is it justifiable and feasible for us to evaluate the special protections offered to them? If so, how?

Robin Lovell-Badge, CBE, FRS, FMedSci, is a senior group leader and head of the Laboratory of Stem Cell Biology and Developmental Genetics at the Francis Crick Institute. Lovell-Badge pointed out that since Jiankui He's presentation at the Second International Summit in 2018, there has been no follow-up or verification on his original data that is available in the public domain. This has made it difficult for any proper medical evaluation of any likely risk. Given that there are people born naturally with CCR5 mutations, including people living in China, these three children should not be considered as 'unique'. It is possible to conduct a genetic test by taking a blood sample to see if the altered CCR5 genes in any of the three children offer HIV protection. Lovell-Badge objected to using the term 'genetic mayhem' to describe the three children. Jiankui He was ethically and scientifically wrong. It was an inappropriate use of a technology that was well acknowledged as not yet safe for clinical applications. However, given that the three children are alive, they may not be that much different from other people, who naturally have mutations. One needs to be careful with using extreme language such as 'genetic mayhem'. Lovell-Badge was of the view that Lulu, Nana and Amy need to be protected, with careful and subtle monitoring of them. But they should be treated like any other child. Their lives should not be invaded by any particular measure.

Helen O'Neill is an expert in reproductive and molecular genetics at the University College London. Helen listed four elements that needed to be addressed. The first was the ethical question of whether germline editing should have been done or should ever be done again. The second was the scientific question on the experimental outcome, the steps that were taken and the choices made on certain gene variants. The third was the personal question of medical misconduct. The fourth was what had happened or might not have happened to the babies themselves. Naturally, these questions are interrelated, but it is important to treat them separately when it comes to discussing what has happened and what will happen in the future. O'Neill emphasised that instead of focusing on what could go wrong, greater attention should be paid to the lessons learned. She agreed that there was no need for genetic exceptionalism to treat the babies differently. What was needed, O'Neill argued, is an ethical sensitivity whereby we feel the need to talk at length about what could, should or might happen, with sufficient scientific evidence and a decent understanding of the context. From a clinical embryology perspective, there is much that we can learn about genome editing from current clinical IVF practice. Scientifically, there is little value in putting these children under special protection plans. More importantly, O'Neill highlighted that we shouldn't automatically assume proper genetic monitoring equals an arduous and burdensome experience. We live in an age where some genetic testing can be done at home with a saliva kit, and monitoring can be achieved through an app, whereby the parents ask their children questions in a non-clinical setting. There is much we could do with minimum intrusion.

Ayo Wahlberg, an expert on the development and routinization of reproductive technologies at the University of Copenhagen, drew on his extensive work on childhood cancer families who have opted to participate in a whole genome sequencing research project (Maeder and Joung 2019; Byrjalsen et al. 2020). 'Surveillance life' is not new healthcare. Wahlberg referred to the example of TP53, a gene whose mutation increases the chance of many cancers. There is a lively debate among bioethicists and geneticists on what kind of surveillance if any should young children enrol in before they turn 18. Clinical conditions such as Lynch syndrome have clearly established mutations with clearly established surveillance programmes. On the question of how children should be informed of their genetic condition, Wahlberg's empirical research (Pedersen et al. 2020) found that when to tell children mattered far less than how the family approached the topic. In the case of the CRISPR babies, it is not yet clear what surveillance is relevant. But how we talk about them is still significant. Wahlberg suggested that, when describing health risks of the three children, approaching genomic questions in terms of known 'clinically actionable mutations' on the one hand, and 'variants of unknown significance (VUS)', on the other, may be more appropriate. He insisted that even when 'surveillance life' is necessary (Heinsen et al. 2022), it remains a family affair and needs to be normalised as much as possible. The conversation that parents and children at some point have needs to be assisted by genetic counselling and informed by empirical research, both when as regards enrolling in/embarking on relevant lifelong surveillance programmes and as regards any questions families might have concerning family planning. Wahlberg also noted that as technology ushered us into new territories, roundtable conversations were a necessary process for us to develop and reflect on ways to talk about things we didn't know how to talk about before.

Discussion

While a key theme of the roundtable discussion was on how individuals and different professions should better understand how to position themselves in relation to genome editing and its social consequences, **Sarah Chan** hoped to see more deliberations on the next steps forward. This pertained not only to China, as three and a half years after the genome-edited babies were born, globally, but we're also still wrestling with how to prepare both the governing structure and the discourses that are essential for future resilience against similar situations. **Robin Lovell-Badge** pointed out that we don't refer to people as products of pre-implantation genetic testing. In a similar vein, labelling these children as CRISPR babies in need of special protection required good reasons, such as the detection of deleterious mutations as a result of the editing.

Sonia Ben Ouagrham-Gormley, associate professor at Schar School of Policy and Government at George Mason University, queried to what extent we can distinguish naturally occurring mutations and those that were manipulated in the lab. This would be useful to clarify whether the children are to be protected against too many intrusions in their private lives. **Helen O'Neill** confirmed that we are already able to tell CRISPR-driven mutations from naturally occurring ones with a biopsy from a saliva sample. **Lovell-Badge** added that possible mosaicism during CRISPR gene editing may mean a single sample could miss or over-represent gene changes. But the technical aspect of genetic follow-up was not intrusive. **Eben Kirksey** noted that, with the broad horizon of possibilities that new technologies open up, the discussion on CRISPR babies has wider implications. That is, how do we live with possible technical accidents and create conditions for people who might be deemed as abnormal by society to flourish? Relatedly, **Jay Johnson** commented that for a long time, his HIV status was in the closet. In the 1980s and early 1990s, HIV was also a highly stigmatised subject in the USA. Johnson, as a practising nurse at the time, was worried about losing his job. When he participated in the Sangamo research and the Associated Press got hold of the story, he needed to 'come out' again to his neighbours and colleagues. However, people embraced him with support and compassion. Johnson concluded that, 'the whole experience of going through the study was interesting, but the reaction I got from my friends and colleagues was amazing'.

Haidan Chen, bioethicist at Peking University, asked **Katie Dow**, Deputy Director of the Reproductive Sociology Research Group at the University of Cambridge, about the lessons that could be drawn from the early phase of the UK media discussion on IVF. Dow has carried out archival research on representations of IVF in the British media, focusing particularly on newspapers and TV documentaries covering the birth of the first 'test-tube baby', Louise Brown (Dow 2019). She noted that when Patrick Steptoe and Robert Edwards established pregnancy in Louise Brown's mother, one of their main fears was that Louise would be born with abnormalities, which would be a catastrophe for the type of science they championed. While the manner of Louise Brown's conception became controversial, at the time of her birth, the UK coverage was almost entirely positive. This was because IVF was seen as a technical assistance to reinforce a normative goal, that is, assisting natural birth resulting from heterosexual marriage. Patrick Steptoe was concerned about the use of IVF and other assisted reproductive technologies beyond heterosexual couples. The human-interest angle that the British press used (Dow 2018), such as depicting the Browns as an ordinary and respectable family, established a subsequent frame for IVF and contributed to its rapid normalisation. In short, Dow reminded the panel that the development of ethical regulation and practices in a country is also a process of self-renewal and reinvention of that nation's projected values on science. She also cautioned about the underlying assumptions that go with describing people as babies and children in English and how it could affect how we perceive responsibility towards them. **Ayo Wahlberg** added that the discussion highlighted the tension between exceptionalism and normalisation. He noted that Jay Johnson, a pioneer in receiving gene-editing treatment, has been through a very exceptional moment in history, but can still go about everyday life as others do. Wahlberg noted that we need to ensure that Lulu, Nana and Amy have access to all the health care they need, but we also need to create a language that helps to 'de-exceptionalise' and to enable them to lead autonomous lives. We need to continue the conversation that Qiu, Lei, and colleagues in China have started, and through the conversation, develop the appropriate language together.

Di Zhang from Peking Union Medical College noted that, to his knowledge, many Chinese bioethicists would concur with Chan's and Hurlbut's views. He underlined the importance of de-exceptionalising these young individuals and supported Ryan's view that the families' needs should take precedence over 'expert opinions'. Di Zhang noted that academics in China have limited influence over top-down decisions, but there is much that can be done to help these families through non-government channels. He also offered an amendment to Eben Kirksey's view. That is, while there are contextual particularities in China's medical morality, the core values embraced by China are also shared universally. More importantly, Di Zhang deemed it important to bear in mind that China's contribution to the global governance of genome editing should not be that of cultural particularity, but should be a better articulation of shared values and shared interests. **Ryan Ferrell** agreed with Di Zhang that while socio-political contexts may vary, culturally, societies may have much more in common than often assumed. He further elaborated on the layers of regulatory failures he witnessed surrounding Jiankui He's research. Ferrell was emphatic about avoiding the practice of naming-and-shaming, but wanted to use this as an opportunity to reflect on why it was difficult to bring visibility to certain social struggles, and on where Chinese bioethicists can exert their influence. A number of participants raised the point of promoting a safe and more enabling environment nationally and transnationally for individuals with concerns or key information to speak out. **Robin Lovell-Badge** emphasised that openness is critical in shaping governance and avoiding future mistakes in this area. He also highlighted that the WHO has already started such work as outlined in the WHO Expert Advisory Committee's Recommendations on human genome editing. This includes working in partnership with a wide range of groups, people, and institutions, within and beyond the global public health community, to identify mechanisms for the effective reporting of illegal, unregistered, unethical, or unsafe research and to promote the best possible open science. **Sonia Ben Ouagrham-Gormley** commented that to some extent, the discussions on Lulu, Nana and Amy have been 'one-way', for they and their family have restricted communications to the outside world on what they desire. **Eben Kirksey** observed that what emerged from the discussion was a genuine interest in pivoting away from blame but towards responsibility. In response to

Di Zhang's comment, Kirksey clarified that his point was not on a West/East divide, but that values are situated in cultural and historical context. Understanding any scientific research and evaluating appropriate responses required taking into account that context. Kirksey believed that there was a general consensus among participants that, firstly, there was still much information that could be shared and secondly, official channels may not be the best (or only) approach to provide care and help for the families. There was a potential for an international group of individuals with different experts to work together to tackle difficult issues.

Ruipeng Lei further explained that their proposal of special protection to the CRISPR babies was not intended to deny them as equals, but to ensure their equity towards developing a good life. She agreed with Robin Lovell-Badge that more transparency and open discussion were required for a scientifically-informed evaluation of the children's needs. She stated that this is what bioethicists are trying to promote in China. The 13 March webinar was the latest effort. **Di Zhang** identified that bioethicists and social scientists in China can take concrete actions to prompt regulatory updates. Part of the reason HIV couples saw Jiankui He's experiment as a solution was because they were excluded from accessing assisted reproductive technologies (ART) in China. When the relevant legislation was made 20 years ago, it made sense as such clinical practice was immature in China and could not safeguard against HIV infection. However, now ART has become safe for HIV couples, but they are still denied these services by outdated regulations. For Di Zhang, the bigger issue is not the three children, but the well-being of the Chinese HIV community in general. **Robin Lovell-Badge** added that there should be flexibility built into regulations that allow a case-by-case examination. The *WHO recommendations on Human Genome Editing* (WHO 2021) suggested the incorporation of diverse governance approaches. In addition, lessons can be learned from the discussions on mitochondrial replacement therapy (MRT) (National Academies of Sciences, Engineering, and Medicine 2016). **Hubert Kang** pointed out that the importance of genetic counselling is often undervalued and misunderstood by the public as well as academics. Proper genetic counselling would have helped desperate patients not to be misled by scientific entrepreneurialism. **Ben Hurlbut** argued that stories that get told by experts shape the life trajectories of people who inhabit those lines, as well as the trajectories of science and ethical apparatus that are used to justify them.

It was a shared view among participants that the roundtable had been highly productive and that the conversation should continue, possibly in a public form.

Additional Chair Communication

Prior to the meeting, the convenor of the roundtable, **Joy Zhang**, communicated her views to participants through emails and direct conversations. Thus, she did not repeat her perspective during the roundtable or discussion but summarised it as follows:

Firstly, bioethicists (and Science and Technology Studies scholars) play a critical role in democratising social and policy discourses on emerging science in China, as is detailed in previous empirical studies (Zhang 2012). This point is often neglected, due partly to the restricted representation of the Chinese bioethics community, which reinforces a common apprehension that Chinese academics are necessarily an extension of the state's political apparatus. In fact, the protection of genome-edited persons has been complicated by bioethicists' own struggle for open and inclusive discussions. There is an imperative for the Global North to recognise that engaging and enabling this community is key to promoting accountable science and innovation in and with China.

However, secondly, the Chinese bioethics community is in dire need of conducting empirically-informed research on the views of scientific practitioners and the general public in China. Indeed, for China's science to establish global influence, it first needs to render 'epistemic justice' to its own people. This pertains not only to social discrimination or power asymmetries in patient-doctor relations. It also pertains to bioethical and scientific communities themselves, who need to expand policy discussions beyond a few elites.

Read in context, the draft proposal's fixation on 'future generations' and on protection were Chinese bioethicists' (misplaced) efforts to correct the very conditions, noted by a few participants, which gave rise to the CRISPR scandal in the first place, such as an exploitative development mentality and a persistent lack of social security. This ironically misplaced good intention further accentuated the necessity for future proposals to be in sync with cultural and technological developments and lived experience.

Finally, panellists pointed out a number of wider implications of the discussion. A further question is whether or not the CRISPR babies protection discussion could be a turning point for Chinese bioethics. In contrast to previously observed 'post-hoc pragmatism' (Zhang 2017), which shuts down debates in the face of global criticism, this time, we've seen commendable efforts from Chinese colleagues to open up the discussion domestically and internationally. Many roundtable participants have pointed out the need for a new 'language' to articulate our relationship with emerging technologies. Surely it is of mutual significance to the world and to China, to have bioethicists from the world's second-largest country for the life sciences (Nature Index 2021), substantively contributing to co-developing a language that is of cosmopolitan resonance.

Key Takeaway Points

- There is a shared objection to genetic exceptionalism as the grounds for protecting genome-edited individuals. Instead, the meeting advocates a sensitive and compassionate approach that prioritises the individuals and their families' wishes, informed by scientific facts.

- The healthcare of genome-edited children should be approached in the same way that the healthcare of all children should be, with the involvement of such specialities as clinical genetics to provide genetic counselling as and when relevant.
- Socially de-exceptionalising the three children is required to enable them to lead normal lives. Empirical research and historical lessons must feed into this de-exceptionalising process.
- Instead of blame, the panel calls for a collective introspection of global scientific and social scientific communities to reflect on the social problems that gave rise to Jiankui He's research and to the decisions of the parents. Concrete actions, such as updating fertility policies and addressing social stigma, can make a meaningful difference to the quality of life of the HIV community in China.
- The panel also calls for the promotion of a safe and more enabling environment nationally and transnationally for individuals with concerns or key information to speak out.
- There is a need to develop more appropriate vocabularies to accurately reflect the nature of the challenge embodied by new technologies and their social consequences.
- Transnational collaborations could help create conditions for gene-edited persons and others in comparable situations to flourish. Multi-disciplinary international conversations such as this one must continue.

List of Participants and Panellists' Biographies

Event Organiser and Chair

Joy Y. Zhang is a sociologist with a first degree in medicine. She is the Founding Director of the Centre for Global Science and Epistemic Justice at the University of Kent. She received her PhD from the London School of Economics and Political Science under the supervision of Sarah Franklin. Her thesis was examined by Ulrich Beck and Christoph Rehmann-Sutter. Her research investigates the transnational governance of scientific uncertainty and the decolonisation of knowledge production. Conceptually, her work contributes to sociological theories of risk, cosmopolitanism, decolonisation, and subaltern politics. She has undertaken empirical studies on stem cells, synthetic biology, genome editing, food movements, and environmental politics. She is the author of four academic monographs: *The Cosmopolitanization of Science: Stem Cell Governance in China* (2012), *Green Politics in China: Environmental Governance and State-Society Relations* (2013), *The Elephant and the Dragon in Contemporary Life Sciences: A Call for Decolonising Global Governance* (2022) and *Democratic Participation and the Cosmopolitics of Science: Why Scientific Citizenship Matters in the twenty-first Century* (2023).

Panellists

Sarah Chan is a Reader in Bioethics at the Usher Institute, University of Edinburgh; she is currently an Associate Director of the Centre for Biomedicine, Self and Society and a Deputy Director of the Mason Institute for Medicine, Life Sciences and Law. Previously, from 2005 to 2015, she was a Research Fellow in Bioethics at the University of Manchester, first at the Centre for Social Ethics and Policy and from 2008 at the Institute for Science Ethics and Innovation. Sarah's research focuses on the ethics of new biomedical technologies, including gene therapy and genetic modification; stem cell and embryo research; reproductive medicine; synthetic biology; and human and animal enhancement. Her current work draws on these interests to explore the ethics of emerging modes of biomedicine at the interface of health care research, medical treatment and consumer medicine, including population-level health and genetic data research; the use of human biomaterials in both research and treatment; and access to experimental treatments and medical innovation. Sarah is a Co-Investigator on the UK Pandemic Ethics Accelerator, a project funded by UKRI that brings together ethical expertise from across the UK to identify and address challenges raised by the Covid pandemic.

Ryan Ferrell was a consultant hired by He Jiankui starting in August 2018. Much of his experience is documented in Eben Kirksey's book: *The Mutant Project*.

J. Benjamin Hurlbut is an Associate Professor in the School of Life Sciences at Arizona State University. He is trained in science and technology studies (STS) with a focus on the history of the modern biomedical and life sciences, and his research lies at the intersection of STS, bioethics, and political theory. He studies the changing relationships between science, politics and law in the governance of biomedical research and innovation, examining the interplay of science and technology with democracy, religious and moral pluralism, and public reason. He is the author of *Experiments in Democracy: Human Embryo Research and the Politics of Bioethics* (2017) and co-editor of *Perfecting Human Futures: Transhuman Visions and Technological Imaginations* (2016), as well as numerous articles and book chapters. He holds an A.B. from Stanford University and a PhD in the History of Science from Harvard University. He was a postdoctoral fellow in the program on Science, Technology and Society at the Harvard Kennedy School.

Jay Johnson is the Director of Volunteers at Action Wellness, a Direct Case Management agency for people living with chronic illness, including HIV. Jay holds a Bachelor of Science degree in Health Administration and an MBA. He is also a retired nurse. He has been living with HIV for 31 years and participated in the small study of the Sangamo BioSciences therapy, the first use in humans of gene editing in 2010. Jay loves travelling, dancing, and spending time with friends and family. He currently lives with his partner of 17 years and their four furry kids (cats) in Philadelphia.

Hubert Kang oversees the operation of the Institutional Review Boards of BGI-Shenzhen and of China National GeneBank. He also lectures at the BGI-College (also known as the BGI Education Center, University of Chinese Academy of

Sciences). His work focuses on promoting effective ethical oversight and ethical reviews, with a particular interest in genomics. He has been part of various national research programmes. Currently, he is part of the National Key Research and Development Program project ‘Synthetic Biology Ethics, Policy and Regulation Framework Research’ and the Ministry of Education-funded project ‘Intellectual Property Issues in Synthetic Biology Development’. He has translated several academic monographs, which have helped to promote cross-cultural idea exchange. He is also a member of the Morality & Ethics Committee of the Chinese Society of Genetics, a member of the Charity and Ethics Committee of the China Health Information and Health Medical Big Data Society.

Eben Kirksey is an American anthropologist who writes about science and justice. He is best known for his pioneering work in multispecies ethnography—an approach to studying human interactions with animals, plants, fungi, and microbes. Duke University Press published his first two books—*Freedom in Entangled Worlds* (2012) and *Emergent Ecologies* (2015)—as well as one edited collection: *The Multispecies Salon* (2014). Recently, he introduced new approaches to chemo-ethnography in collaboration with Nicholas Shapiro. Currently, he is an Associate Professor (Research) at Alfred Deakin Institute in Melbourne, Australia. The Institute for Advanced Study in Princeton, New Jersey, hosted Kirksey for the 2019–2020 academic year, where he finished his latest book: *The Mutant Project*.

Ruipeng Lei is Professor of Bioethics, Executive Director of the Centre for Bioethics, School of Philosophy, Huazhong University of Science and Technology, China. She is the Vice President and Secretary-General of the Chinese Society for Bioethics and a Board Member of the Chinese Society for Ethics. She has researched and advised the Chinese government on regulatory issues regarding organ transplant, human genetic resource, synthetic biology, genome editing and biobanking. She is the PI of China’s National Key Research and Development Program, ‘the Ethical and Policy Framework of Synthetic Biology’, and the PI of National Major Social Science Grant, ‘Philosophical Investigation of BioBanking in the Era of Big Data’. She was a Research Fellow at Harvard School of Public Health, 2003–2005. She is a Hastings Center Fellow and a member of the WHO COVID-19 Ethics and Governance Working Group.

Robin Lovell-Badge, CBE, FRS, FMedSci, is a senior group leader and head of the Laboratory of Stem Cell Biology and Developmental Genetics at the Francis Crick Institute.

He obtained his PhD in embryology at University College London (UCL) in 1978 with Martin Evans. After postdoctoral research in the Genetics Department at the University of Cambridge and at the Institut Jacques Monod in Paris, Robin established his independent laboratory in 1982 at the Medical Research Council (MRC) Mammalian Development Unit, UCL, directed by Anne McLaren. In 1988, he moved to the MRC National Institute for Medical Research, which was incorporated into the Francis Crick Institute in 2016.

Robin is well known for his discovery of *Sry*, the Y-linked testis-determining gene, and for subsequent work that established pathways involved in the initiation and maintenance of gonadal sex. Robin’s lab also discovered the *Sox* gene family, revealing that *Sox2* is important for pluripotency in the early embryo and how several *Sox* genes are critical for the development of the central nervous system, the pituitary, and for stem cells in these systems. In addition to its fundamental interest, Robin's work is of clinical relevance.

Robin is also active in public engagement and policy work, notably around stem cells, genetics, human embryo and animal research, and in ways science is regulated and disseminated. For example, he has served on committees for the Human Fertilisation and Embryology Authority, contributing to changes in the UK laws around human embryo research, and for the Royal Society, US National Academies of Science, and WHO, exploring applications of human genome editing and its governance. He has received the Louis Jeantet Prize for Medicine (1995), the Amory Prize (1996), the Feldberg Foundation Prize (2008), the Waddington Medal of the British Society for Developmental Biology (2010), the ISSCR Public Service Award (2021), and the Genetics Society Medal (2022).

Helen O'Neill is a lecturer in Reproductive and Molecular Genetics and Director for the MSc in Reproductive Science and Women's Health at the Institute for Women's Health, University College London (UCL). Her research focuses on preimplantation embryo development and the use of genome editing to assess understanding and treatment of disorders of infertility. Dr. O'Neill has an honours degree in Genetics, an MSc in Prenatal Genetics and Fetal Medicine from UCL, and did her PhD and postdoctoral research on the genetics of ovarian development in the Department of Stem Cell Biology and developmental genetics at the National Institute for Medical Research. She lectures both master's and medical students and is active in public engagement involving genome editing. Dr. O'Neill is CEO and Founder of Hertility Health, a precision medicine-based approach to reproductive health and fertility.

Ayo Wahlberg is Professor MSO at the Department of Anthropology, University of Copenhagen. His research has focused on the development and routinization of reproductive technologies in China and beyond. Ayo is the author of *Good Quality—the Routinization of Sperm Banking in China* and co-editor of *Selective Reproduction in the 21st Century*.

Discussion Participants

Sonia Ben Ouagrham-Gormley—Associate Professor, Schar School of Policy and Government at George Mason University
Haidan Chen—Associate Professor, School of Health Humanities, Peking University
Katie Dow—Deputy Director, Reproductive Sociology Research Group, University of Cambridge
Lucy Lu Gao—Associate Professor, Institute for the History of Natural Sciences, Chinese Academy of Sciences
Di Zhang—Associate Professor, Peking Union Medical College

Observer

Smriti Mallapaty—Senior Reporter for the Asia-Pacific region, *Nature*

References

Business Wire. 2021. EdiGene announces approval of its IND application for its CRISPR/Cas9 gene-editing hematopoietic stem cell therapy ET-01 in β-thalassemia by China national medical products administration. *Business Wire*, January 18, 2021. https://www.businesswire.com/news/home/20210118005210/en/EdiGene-Announces-Approval-of-its-IND-Application-for-its-CRISPRCas-9-Gene-editing-Hematopoietic-Stem-Cell-Therapy-ET-01-in%2D-%2Dthalassemia-by-China-National-Medical-Products-Administratio.

Byrjalsen, Anna, Ulrik Kristoffer Stoltze, Anders Castor, and Ayo Wahlberg. 2020. Germline whole genome sequencing in pediatric oncology in Denmark: Practitioner perspectives. *Molecular Genetics & Genomic Medicine* 8 (8): e1276. https://doi.org/10.1002/mgg3.1276.

Cyranoski, David. 2019. China set to introduce gene-editing regulation following CRISPR-baby furore. *Nature*, May 20, 2019. https://doi.org/10.1038/d41586-019-01580-1.

Dow, Katharine. 2018. 'Now she's just an ordinary baby': The birth of IVF in the British press. *Sociology* 53 (2): 314–329. https://doi.org/10.1177/0038038518757953.

Dow, Katharine. 2019. Looking into the test tube: The birth of IVF on British television. *Medical History* 63 (2): 219–236. https://doi.org/10.1017/mdh.2019.6.

Heinsen, Laura Louise, Ayo Wahlberg, and Helle Vendel Petersen. 2022. Surveillance life and the shaping of 'genetically at risk' chronicities in Denmark. *Anthropology & Medicine* 29 (1): 29–44. https://doi.org/10.1080/13648470.2021.1893654.

Jia, Hepeng. 2019. China approves ethics advisory group after CRISPR-babies scandal. *Nature*, August 8, 2019. https://doi.org/10.1038/d41586-019-02362-5.

Kirksey, Eben. 2020. *The mutant project: Inside the global race to genetically modify humans*. New York: St. Martin's Press.

Lew, Linda. 2021. Legal experts urge people to take new Chinese biosecurity law seriously. *South China Morning Post*, April 16, 2021. https://www.scmp.com/news/china/science/article/3129740/legal-experts-urge-people-take-new-chinese-biosecurity-law.

Maeder, Morgan L., and Keith A. Joung. 2019. Genome editing technologies for gene and cell therapy. *Molecular Therapy* 27 (3): 461–465. https://doi.org/10.1016/j.ymthe.2018.07.001.

Mallapaty, Smriti. 2022. How to protect the first 'CRISPR babies' prompts ethical debate. *Nature* 603:213–214. https://doi.org/10.1038/d41586-022-00512-w.

National Academies of Sciences, Engineering, and Medicine. 2016. *Mitochondrial replacement techniques: Ethical, social, and policy considerations*. Washington, DC: National Academies Press. https://www.nationalacademies.org/publications/21871.

Pedersen, Line Hjøllund, Ayo Wahlberg, Marie Cordt, Kjeld Schmiegelow, Susanne Oksbjerg Dalton, and Hanne Bækgaard Larsen. 2020. Parents' perspectives of the pathway to diagnosis of childhood cancer: A matter of diagnostic triage. *BMC Health Services Research* 20:969. https://doi.org/10.1186/s12913-020-05821-2.

Shenzhen Municipal People's Government. 2020. *Announcement on the approval and issuance of Shenzhen local standards (No. 34) (深圳市地方标准批准发布公告，总第34号)*. Shenzhen Government Online. https://www.sz.gov.cn/szzt2010/wgkzl/jcgk/jcygk/zdzcjc/content/post_8158235.html.

Song, Lingqiao, and Yann Joly. 2021. After he Jiankui: China's biotechnology regulation reforms. *Medical Law International* 21 (2): 174–191. https://doi.org/10.1177/0968533221993504.

Tebas, Pablo, David Stein, Winson W. Tang, Ian Frank, Shelley Q. Wang, S. Gary Lee, Kaye Spratt, et al. 2014. Gene editing of CCR5 in autologous CD4 T cells of persons infected with HIV. *New England Journal of Medicine* 370 (10): 901–910. https://doi.org/10.1056/NEJMoa1300662.

World Health Organization. 2021. *Human genome editing: Recommendations* (Geneva: World Health Organization, 2021). https://www.who.int/publications/i/item/978924003038.

Zhang, J. Y. 2012. *The cosmopolitanization of science: Stem cell governance in China*. Basingstoke: Palgrave Macmillan.

Zhang, J. Y. 2017. Lost in translation? Accountability and governance of clinical stem cell research in China. *Regenerative Medicine*. 12 (6): 647–656. https://doi.org/10.2217/rme-2017-0035.

Joy Y. Zhang is Professor of Sociology with a first degree in medicine. She is also the Founding Director of the Centre for Global Science and Epistemic Justice (GSEJ) at the University of Kent. An internationally recognised expert in the transnational governance of scientific risk, she specialises in emerging life sciences that impact both human and environmental futures. Her work advances sociological thinking on risk, cosmopolitanism, decolonisation, and subaltern politics, with a comparative focus on four major scientific powers: China, India, the UK, and the US.

Zhang is especially known for her expertise in science diplomacy: how scientific collaboration, governance, and public engagement can serve as sites for international negotiation, trust-building, and ethical coordination in times of geopolitical tension. In this regard, she has been at the forefront of efforts to connect responsible innovation, global governance, and diplomatic practice in the life sciences. She has initiated several landmark programmes, including the BioGovernance Commons (2021), the founding of GSEJ (2022)—whose research has informed G20 and G7 policy discussions—and, most recently, O.D.E.SS.I., a global 'odyssey' for public engagement with the life sciences. She is the author of three academic books and publishes widely in high-impact journals across both the natural and social sciences. She is frequently interviewed by global media and advises leading research and policy institutions in Europe and Asia.

Chapter 13
APPENDIX 2: Looking Back into the Future: CRISPR and Social Values (2023)

Joy Y. Zhang

On 11 February 2023, the Centre for Global Science and Epistemic Justice at the University of Kent hosted an open discussion on CRISPR technology for and with

---

Editorial note: This piece documents a second major international discussion addressing an ethically controversial and politically sensitive issue concerning the governance of human genome editing among Chinese stakeholders and global counterparts. Notably, it marked the first occasion on which Dr. Jiankui He, the scientist widely known for breaching international ethical norms by conducting heritable human genome editing in 2017, agreed to participate in a public exchange with Chinese bioethicists and genome-editing researchers. Equally significant, the session created space to hear directly from a broader range of Chinese practitioners working in CRISPR research, offering insights into their ambitions, concerns, and perspectives.

The discussion was aligned with the BioGovernance Commons' commitment to promoting responsible science and fostering inclusive social and policy dialogue. Rather than aiming for consensus, the event sought to encourage greater openness and engagement among Chinese scientists and bioethicists, both domestically and internationally, on this complex issue. Lasting over 2 h, the event brought together approximately 110 participants from 13 countries, including representation from at least 12 of China's provincial-level jurisdictions, with around 70 attendees based in China.

The event received wide media coverage, including features by BBC *Science in Action* and *Inside Science*, as well as reporting from outlets such as ABC Radio National (Australia), the Associated Press, BBC News, *Nature*, *Stat News*, *South China Morning Post*, *Global Times*, *China's Science and Technology Daily*, and Yahoo News Japan. This report was originally published online on 13 February 2023 by the Centre for Global Science and Epistemic Justice at the University of Kent. The views expressed reflect those of the participants and do not represent the positions of any affiliated institutions. The original report should be cited as: Joy Y. Zhang, Ruipeng Lei, Kathleen Vogel, Sonia Ben Ouagrham-Gormley (2023) Looking Back into the Future: CRISPR and Social Values: BioGovernance Common's Open Discussion with Chinese Academics. 13 February. Canterbury: Centre for Global Science and Epistemic Justice, University of Kent

---

J. Y. Zhang (✉)
Centre for Global Science and Epistemic Justice, University of Kent, Canterbury, UK
e-mail: y.zhang-203@kent.ac.uk

J. Y. Zhang et al. (eds.), *BioGovernance Commons and China's Role in Global Risk Governance*, Emerging Globalities and Civilizational Perspectives,
https://doi.org/10.1007/978-3-032-32198-5_13

Chinese academics, titled 'Looking Back into the Future: CRISPR and Social Values'. This was part of the BioGovernance Commons initiative founded by us in the summer of 2021, which functions as a trusted forum for researchers inside and outside China to meet regularly, under the motto 'sharing perspectives on shared challenges.' The event lasted more than 2 h and attracted participants from 13 different countries, as well as academics from at least 12 of the 34 provincial jurisdictions in China. In total, approximately 110 people attended the meeting, 70 of whom were based in China.

Executive Summary

As organisers, we highlight these main points from the event:

1. We've exposed that there is little substance behind the ambitious Duchenne muscular dystrophy (DMD) gene therapy that Dr. Jiankui He, the scientist behind the illegal heritable human genome editing, has been boasting about to global patients. We were concerned that he might endanger another vulnerable population if his new venture remains unchecked.
2. Our event nudged the Chinese media to publicly challenge Dr. He for the first time on his DMD research. As a result Dr. He has announced that he will prioritise research rather than social media promotion (Sun 2023).
3. The discussion exposed regulatory and communicative gaps in China, which we hope will receive due attention from relevant authorities. It became apparent in the discussion that there was no clarity about what type of restrictions Chinese authorities had imposed on Dr. He's research after the CRISPR babies controversy, and what institutions were responsible for enforcing these restrictions. Chinese participants have made explicit calls for more regulatory clarity within China.
4. Three university CRISPR scientists, Prof. Shihui Yang, Prof. Li Su and Prof. Shangxian Xie, shared their diverse views on how bioethical guidelines function (or fail to function) in everyday research in China.
5. As discussants, bioethicist Prof. Ruipeng Lei and legal scholar Prof. Jiayou Shi provided a brief review on how bioethical and legislative discussions have shifted in China since 2018 and speculated on future regulations.
6. 'Can the Chinese scientists speak?' Similar to our previous events (Zhang et al. 2022), we noticed that Western media bias and Chinese censorship habitually 'mute' Chinese scientists. We believe the best way to demystify 'Chinese science' is to enable more ordinary scientists to have a voice in both domestic and international debates. We hope both the strengths and limits of our event can inspire more colleagues to improve the ways we promote critical yet constructive international dialogues.

Our Main Aim

We convened this session because China is an important player in the responsible development of gene technologies globally, yet how bioethics *actually* functions in everyday innovation in China may remain elusive. More importantly, little is known about the state of academic discussion on CRISPR genome editing in China and Chinese scientists' views on this. This is why we invited three Chinese CRISPR scientists at different stages of their careers and with varied experience in policy or public engagement. They were Professor Li Su, Professor Shihui Yang, and Professor Shangxian Xie.

Another factor that prompted us to organise this session is that we wanted to hear more voices from ordinary Chinese scientists, and not just a few elites. We were fully aware that practical constraints and institutional censorship prevented us from hearing more diverse voices from China. For many (if not the majority of) Chinese academics, speaking to Western media or speaking publicly in a domestic setting often requires lengthy institutional approval procedures. Participating in academic events is one of the few exceptions where Chinese colleagues can express their views publicly and exchange ideas freely with an international audience. In line with BioGovernance Commons' commitment to promote inclusive and democratic discussions, we wanted this event to enable Chinese voices from different levels and from different disciplines. To facilitate the participation of Chinese academics, we decided to use Chinese as the main language, with simultaneous translation in English.

Our main aim, as was stated in our event announcement (GSEJ 2023) and in communications sent to participants, has always been to encourage more Chinese scientists and bioethicists to engage in open dialogue on this important topic both in China and internationally.

Why We Invited Dr. Jiankui He

During the planning of this event, it came to our attention that due to political sensitivity, China has not yet had an open discussion on CRISPR technology and ethical governance with Dr. Jiankui He since he announced his illegal experiment with heritable human genome editing in 2018 (Marchione 2018), for which he was sentenced to 3 years in prison. However, since being released, Dr. He has resumed genome research. He has publicly launched an ambitious plan for his current gene therapy research on Duchenne muscular dystrophy (DMD) to enter clinical trials by March 2025 and has been actively recruiting patients. As Dr. He has effectively become a taboo topic within China, most people avoid discussing him in public. Thus, a pervasive side effect of Chinese censorship is that it has left Dr. He considerable freedom to self-promote on social media, leaving his claims unchecked and unchallenged. In planning this event, we've consulted a number of key regulators and academics, and have noticed a shared, yet unspoken, concern that there is no clarity or transparency over how Dr. He's research is held accountable.

We consider it important for Chinese academia to be able to discuss controversial cases in public. We also consider it critical to have a frank conversation about Dr. He's current research to understand how accountability is established and to ensure that the exploitation of desperate patients and vulnerable individuals will not happen again.

Thus, similar to the other three speakers, we invited Dr. He to give a talk addressing the theme of the seminar 'Looking Back into the Future'. That is, we've asked all speakers to *look back* at their research experience and comment on how ethics and scientific research can be better integrated and shed light *on the future*, on what they see as the types of dialogues or actions needed.

Dr. He accepted our invitation in December 2022. Seventeen hours before the event, we, the organisers, learnt from Dr. He's Twitter account that he was 'not ready to talk about [his] experience in the past 3 years.' We were deeply disappointed by this. But we thought it would still be a valuable opportunity to provide clarity on his ongoing research with DMD patients and what has been put in place to ensure such ambitions will be carried out responsibly and ethically.

We also anticipated that Dr. He may use this event as a publicity stunt and may refuse to engage with critical questions. However, we considered that, given his active recruitment campaign of DMD patients, having his refusals to clarify key questions recorded in a public event was also of public interest. We made clear at the beginning of the event that what is said and what is left unsaid in the meeting are equally informative.

Our Approach to the Media

Given that Dr. Jiankui He had used the Harvard closed-door event and an invitation from the University of Oxford to create the false illusion of academic endorsement, we took extra care of not having this public event hijacked by any individual's public relation narrative.

Since the *Guardian*'s interview with Dr. He was published on 4 February (Devlin 2023), we've received many media requests around the world to attend this event.

We needed to balance transparency and fairness, as well as the practical concerns that the presence of global media may discourage many Chinese participants from attending. The approach all four convenors (Ben Ouagrham-Gormley, Lei, Vogel and Zhang) of the BioGovernance Commons initiative agreed on was to only invite academic journals and Chinese scholarly media. We thus declined the request for attendance from all other media.

As the aim is to promote openness in and with China, we did not impose the Chatham House rule. In fact, we were very clear in communications with all participants that selected scholarly media would be present. We further reminded participants of this point during the event.

Participants were also informed that we recorded the event only for note-taking purposes and that the video would not be shared. Our intention to publish a summary of the event so that the main points of discussion would be made public knowledge was also made clear before the event.

A Summary of the Speakers' Input

Dr. Jiankui He delivered the opening talk. Although we asked the other three speakers to give 15-min presentations, we gave Dr. He a 40-min slot. This was in recognition of the complicated history of his research, and we wanted to give him sufficient space to share and elaborate on his plans and his views. At the event, Dr. He's presentation 'CRISPR Gene Editing: How Revolutionary Biotechnologies Make Our Lives Better' lasted only 25 min, which mostly consisted of explaining basic scientific knowledge, such as the structure of DNA. He spent much time paying tribute to other well-celebrated major scientific advancements and breakthroughs, before slipping in his own research, of which Dr. He only talked for 1 min and 50 s. This included confirming that he aimed to raise 50 million Chinese RMB through charity and enter a clinical trial by March 2025. He mentioned the plan of setting up an international ethics committee for his current lab, but did not give any specification on its membership, plan, or timeline. He was of the view that society was supportive of using gene editing to cure disease and that the regulations in both the USA and China should be made clearer. At the end of the presentation, Dr. He read out the presentation slide that 'heritable embryo gene editing is not allowed in human clinical practice, either in China or other countries'. When later pressed by a professor from Shanghai Jiaotong University if he thought that heritable human genome editing should be 'banned', Dr. He declined to comment.

Prof. Shihui Yang from the School of Life Sciences, Hubei University, was the deputy head of the State Key Laboratory of Biocatalysis and Enzyme Engineering. CRISPR technology in his research on industrial microbiology. He noted that life scientists' engagement with bioethics cannot be taken for granted. His interest in engaging with ethics discussions was mostly propelled by his founding of the Hubei Provincial Association for Synthetic Biology. He noted that it was scientific journals in China that put him in contact with bioethicists and helped him to get access to social science events. Through his teaching experience both in China and in the USA, Prof. Yang pointed out that drawing science students' attention to biosecurity and bioethics debates could be a common challenge in university education, and should be seen as important as developing cutting-edge technologies. To improve bioethical literacy among Chinese scientists, Prof. Yang suggested that a stronger presence of bioethical discussions at academic conferences and international competitions, such as iGEM, may be a more effective way. He also pointed out the important role international collaboration plays in promoting bioethical awareness. His final slide summarised his talking points as 'ABCED': Attention, Balance, Collaboration, Discussion, Education.

Prof. Li Su from Huazhong University of Science and Technology uses CRISPR in her cancer research. She agreed with Dr. Jiankui He's view that any life scientist who doesn't know how to apply CRISPR to their research is outdated. While CRISPR may not always be a 'necessary' tool, it is an effective one. Her slides showed past work her team have published on how biosecurity can be better managed in China and on an innovative approach they've experimented with undergraduate training of biosecurity (Zeng et al. 2022). Her take on bioethics is rooted in her commitment to promoting 'One Health', which is a collaborative, multisectoral, and transdisciplinary approach to promote sustainable human and environmental wellbeing. But Prof. Su also pointed out an often ignored aspect, that is, there is an equal need for social science students to understand avant-garde science. This was the rationale for her to introduce the module 'Modern Life Science Experiment' specifically for social science students. This point was later echoed by Prof. Shihui Yang when answering a question at the end of the seminar. Prof Yang pointed out a frustration that there have been research findings in China that cannot be put into application due to broad regulations. He called for more scientific literacy among policy makers and regulators, so that Chinese regulators can have a more differentiated approach that recognises varied applications and development stages of an innovation.

Prof. Shangxian Xie was also from Huazhong University of Science and Technology and belongs to a younger generation of Chinese scientists. He uses CRISPR in his microbiological and animal research. He recounted his misconception of bioethics during his student years, in which he thought ethics were limited to conceptual debates on morality and values. It was when he started his PhD study in the USA, and when he was asked to sign forms on bioethics and biosecurity before starting to work in the lab, that he realised ethics had concrete legal implications too. He pointed out that different sub-fields within science have their own ethical norms and language. While it is not unusual for scientists to move between research topics or run projects in different sub-areas at the same time, it cannot be taken for granted that scientists necessarily know what appropriate ethical procedures they should comply with. One personal anecdote he shared was that as a scientist who mainly worked with microorganisms, he was well aware of what type of environmental assessments and protection procedures were needed. So, when he applied for a national grant monitoring changes in mice faeces in relation to microorganism intake, he did not know a separate ethics review for animal-related research was needed until the university's research office rejected his proposal. He pointed out that regional disparity in ethics assessment and enforcement standards remains a challenge in China. Prof. Xie emphasised both in his talk and in his answer to participants' questions the intrinsic conflicts between science and ethics. He underlined the explorative and future-oriented nature of science, and contrasted this with the restrictive and history-based characteristics of bioethics. How to balance these two conflicting forces, for Prof. Xie, was the key challenge. He suggested that this should be a key task for BioGovernance Commons as the initiative brings together interdisciplinary and intercultural input.

A Summary of the Discussion

We had approximately 40 min for discussion. Approximately two-thirds of the questions were asked by Chinese participants. Given the lack of substance of Dr. Jiankui He's presentation and opaqueness of his 2018 scandal, not surprisingly, most of the questions were directed to him. We were disappointed that Dr. He refused to engage in any discussion beyond his prepared talk. Below is a list of questions that were asked by participants based in and outside of China:

- Are you restricted from doing any type of research in China?
- Are you restricted from speaking about the research you did 3 years ago, either by the Chinese government, your former university, or your family?
- What kind of concerns do you have moving forward?
- What are your thoughts on communicating with media, rather than bioethicists or other academic peers, about your research?
- Could you please say more about your current DMD research procedure? Given this is a disease which may benefit from early interventions, what is your planned age range of patients?
- Are you restricting your DMD research to somatic gene therapy, or does it incorporate germline research?
- What's your plan for making your gene therapy accessible to patients?
- What's your view on whether human germline editing should be banned?
- What's your view on China's subsequent regulations in genome editing? Do you think it has become more relaxed or more tightened?
- It is expected that China's Ministry of Science and Technology and the National Health Commission will promulgate even more guidelines and regulations in this area. Do you think this is a good thing or a bad thing? Or do you have any comments on this?
- You've mentioned in your talk that China's governance is not yet comprehensive. Could you say a bit more about where the regulatory gaps are and how you think these gaps can be addressed?
- What shapes your understanding of bioethical concerns? Is it through university education, your training abroad, research culture in China or something else?

To all the questions asked, Dr. Jiankui He refused to answer but asked participants to send him their questions by email. Despite the lack of response from Dr. He, Chinese participants engaged in the discussion and expressed diverse views. A professor from Shenzhen expressed the view that the Chinese authorities' penalty on Dr. He was too harsh. In his view, there has been an aggravated antagonism between ethicists and scientists in China. He referred to David Baltimore's 1975 Asilomar Conference on recombinant DNA as a model in leading bioethics discussions. This reference provided a stark contrast to Dr. Jiankui He's refusal to engage in the seminar. Those who are familiar with China's life sciences may recall that the early contemporary development of Chinese bioethics and biogovernance was scientist-led. Huizhen Sheng, who worked on human-rabbit hybrid embryos, and

Huanming Yang, who led China's participation in the Human Genome Project, are two key examples of scientists who led the discussion about bioethics.

Weiwei Cao from Shanghai University was 'upset' that Dr. He refused to answer any questions. She said while she admired his venturesome courage, she was disappointed that He made little comment on social values and failed to mention any possible negative implications of his research on human subjects. She was deeply disappointed that neither the time in prison nor the bioethical and legal guidance from the wider Chinese academic community had made him take ethical and social issues more seriously. She asked him to consider how one small step for one scientist, such as his 2018 heritable human genome editing experiment, could create significant backlash to his peers and stall the field. Cao was of the view that contemporary science cannot be developed behind closed doors. China needs to engage with global dialogues.

Lucy Gao from the Chinese Academy of Science said she was able to feel some sincerity from Dr. He, as he did agree to talk about his work. But she was equally frustrated that nothing of substance was shared 'I think he does want to connect with the wider world and have better exchange,' said Gao, 'so I invite him to make more effort in addressing questions raised by academics and the public.

There were also two longer inputs from Chinese colleagues.

Prof. Ruipeng Lei, who co-chaired the event, provided a brief review of China's bioethics discussions on genome editing, chiefly from her personal work experience. She noted a series of key events. This included (1) the 8th National Bioethical Annual Conference hosted by Fudan University between 16 and 18 November 2018. Prof. Lei chaired a session on the potential abusive use of genome editing. This was also the first time she met Jiankui He, (2) the adoption of 'ethical governance' into China's national agenda, through the Central government's founding of the National Science and Technology Ethics Committee (Sun 2020). (3) A series of domestic discussions on the protection of genome-edited individuals between January and March 2022: This included a joint bioethics discussion held jointly by Huazhong University of Science and Technology, Renmin University, and Xiamen University on 13 March 2022. Approximately 138 participants joined the meeting. That was China's first open discussion on the bioethics of supporting individuals born through heritable genome editing. This national meeting was prompted by the 18 March 2022 International Roundtable (Centre for Global Science and Epistemic Justice 2022) that was convened by the Centre for Global Science and Epistemic Justice at the University of Kent, which demanded transparency and democratic deliberation on this issue. (4) In March 2022, China's central government promulgated the *Guidelines on Enhancing Ethical Governance of Science and Technology* (Central Committee of the Communist Party of China 2022). This has led to the establishment of a dedicated task force on university education on ethics in science and technology. (5) In addition to a revised version of the *Implementing Rules for Human Genetic Resources Management*, Lei also speculated that a revised and updated version of the *Measures for the Ethical Review of Biomedical Research Involving Humans* will also be published soon. It is expected to be wide in scope, incorporating all research involving human material or human data. It is also expected to impose similar requirements on ethical committees and ethical review procedures in hospitals, universities and other research institutes alike.

Jiayou Shi, Professor of Law at Renmin University, took part in writing Article 1009 of China's Civil Code in 2020, which specified that all clinical and scientific research that is related to the human genome and human embryos has to comply with relevant legislation, administrative rules and guidelines. In 2020, China's Criminal Law also expanded its Article 336, which prohibits reproductive genome editing and human cloning. Prof. Shi also drew attention to an increasing legislative attention to inter-generational fairness in China. Prof. Shi further commented on Dr. Jiankui He's remark in his recent Guardian interview that he acted 'too quickly' (Devlin 2023). Shi pointed out that there were different ways to comprehend what was 'too quick': It could be a technical pre-maturity, in which a technology was applied before its safety and effectiveness were tested. But Shi considered Dr. He acted 'too quickly' because he took action before giving his action due ethical consideration.

Looking Back into the Future: The BioGovernance Commons' Perspective

Looking back at the organisation, delivery, and initial perception of this seminar, we found ourselves at the crossfire of Western biases against Chinese science and China's intensified censorship. We had similar experiences with our previous events on biosecurity and on supporting the individuals born from heritable genome editing.

Since Jiankui He publicised our event through his Guardian interview, we have received unfounded criticisms from the media and academic peers of rehabilitating He's reputation, whitewashing China and tarnishing the reputation of the field of bioethics and biogovernance. Our main aim of encouraging more Chinese academics to engage in open dialogue on this important topic, both in China and internationally, was repeatedly brushed aside by these criticisms.

For individuals and institutions based in China, to host an open discussion on controversial issues is often extremely challenging, as concerns over social stability often restrain discussion. The political sensitivity of the event's topic also demands a neutral platform. The BioGovernance Commons initiative is honoured to be a host and enabler for such an important debate.

We are conscious that participants of this seminar may consider the 'worthiness' of attending such an event differently. Some considered that its worthiness lies in knowing more about Jiankui He's past and present scandals, some wanted to know the inner working of Chinese science and Chinese politics, some were curious to learn global peers' views. Depending on what one hoped to get out of the seminar, participants may have left the event feeling informed, enlightened, puzzled, or disappointed.

As organisers, we consider all the efforts worthwhile, as our 150-min event has helped minimise the chance of another scandal exploiting desperate patients and vulnerable individuals. We believe that Chinese scientists' calls for regulatory clarity at the event are more meaningful than commentaries from global observers. We are delighted that the event has enabled Chinese media to publicly question Jiankui He's research and alerted Chinese authorities to regulatory gaps.

We highlight that avoidance of engaging with controversial cases and individuals may effectively leave questionable research projects unchecked and unchallenged. Neither should investigations into controversial cases be shrouded in secrecy, nor should access to information be a privilege of a small circle of (political or academic) elites. We hope both the strengths and limits of our event can inspire more colleagues to improve the ways in which we promote critical yet constructive international dialogue.

References

Central Committee of the Communist Party of China and State Council of the People's Republic of China. 2022. *Opinions on strengthening the governance of science and technology ethics issued by the general office of the CPC central committee and the general office of the state council (中共中央办公厅 国务院办公厅印发《关于加强科技伦理治理的意见).* China Government Network, March 20. https://www.gov.cn/zhengce/2022-03/20/content_5680105.htm.

Centre for Global Science and Epistemic Justice. 2022. *The ethics of protecting 'CRISPR babies': An international roundtable summary report*. April 2022, Canterbury: University of Kent.

Devlin, Hannah. 2023. Scientist who edited babies' genes says he acted 'too quickly'. *The Guardian*, February 4, 2023. https://www.theguardian.com/science/2023/feb/04/scientist-edited-babies-genes-acted-too-quickly-he-jiankui.

Marchione, Marilynn, 2018. AP exclusive: Chinese researcher claims first gene-edited babies. *Associated Press (AP News)*, November 26, 2018. https://apnews.com/article/ap-top-news-international-news-ca-state-wire-genetic-frontiers-health-4997bb7aa36c45449b488e19ac83e86d.

Sun, Zifa. 2020. China has established a national science and technology ethics committee (中国已成立国家科技伦理委员会). *China News Service*, October 21, 2020. https://www.chinanews.com.cn/gn/2020/10-21/9319022.shtm.

Sun, Yu. 2023. He Jiankui reappears after gene-edited babies incident, avoids discussing research ethics (基因编辑婴儿事件后贺建奎再现身，对科研伦理避而不谈). *Science and Technology Daily (China)*, February 13. https://mp.weixin.qq.com/s/-ufk666In7Gvjn7HXgLZhQ.

Zeng, Xiaomei, Hailun Jiang, Guangying Yang, Ou Yakun, Lu Shan, Jia Jiang, Ruipeng Lei, and Su Li. 2022. Regulation and management of the biosecurity for synthetic biology. *Synthetic and Systems Biotechnology* 7 (2): 784–790. https://doi.org/10.1016/j.synbio.2022.03.005.

Zhang, Joy Y., Sonia Ben Ouagrham-Gormley, and Kathleen M. Vogel. 2022. Creating common ground with Chinese researchers. *Issues in Science and Technology* 38 (4): 45–48. https://issues.org/common-ground-chinese-researchers-biogovernance/.

Joy Y. Zhang is Professor of Sociology with a first degree in medicine. She is also the Founding Director of the Centre for Global Science and Epistemic Justice (GSEJ) at the University of Kent. An internationally recognised expert in the transnational governance of scientific risk, she specialises in emerging life sciences that impact both human and environmental futures. Her work advances sociological thinking on risk, cosmopolitanism, decolonisation, and subaltern politics, with a comparative focus on four major scientific powers: China, India, the UK, and the US.

Zhang is especially known for her expertise in science diplomacy: how scientific collaboration, governance, and public engagement can serve as sites for international negotiation, trust-building, and ethical coordination in times of geopolitical tension. In this regard, she has been at the forefront of efforts to connect responsible innovation, global governance, and diplomatic practice in the life sciences. She has initiated several landmark programmes, including the BioGovernance Commons (2021), the founding of GSEJ (2022)—whose research has informed G20 and G7 policy discussions—and, most recently, O.D.E.SS.I., a global 'odyssey' for public engagement with the life sciences. She is the author of three academic books and publishes widely in high-impact journals across both the natural and social sciences. She is frequently interviewed by global media and advises leading research and policy institutions in Europe and Asia.

GPSR Compliance

The European Union's (EU) General Product Safety Regulation (GPSR) is a set of rules that requires consumer products to be safe and our obligations to ensure this.

If you have any concerns about our products, you can contact us on ProductSafety@springernature.com

In case Publisher is established outside the EU, the EU authorized representative is:

Springer Nature Customer Service Center GmbH
Europaplatz 3
69115 Heidelberg, Germany

Batch number: 10823959

Printed by Printforce, the Netherlands